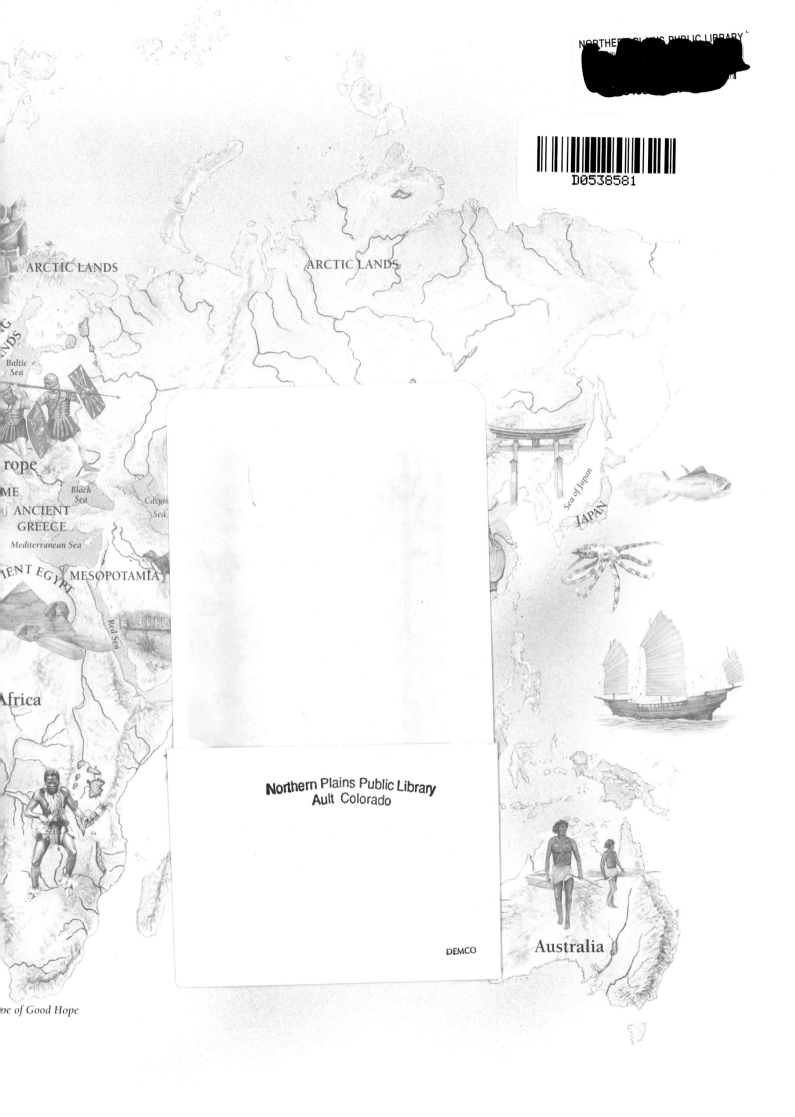

ARCTIC LANDS

ARCTIC LANDS

Baltic
Sea

rope

ME

ANCIENT
GREECE

Black
Sea

Caspia
Sea

Mediterranean Sea

ENT EGYPT

MESOPOTAMIA

Red Sea

Africa

Sea of Japan

JAPAN

Australia

e of Good Hope

SCIENCE, CRAFTS & TECHNOLOGY
through the ages

Series Editor Dr. John Haywood

LORENZ BOOKS

First published by Lorenz Books in 2001

© Anness Publishing Limited 2001

Published in the USA by Lorenz Books
Anness Publishing Inc.
27 West 20th Street
New York
NY 10011

Lorenz Books is an imprint of Anness Publishing Inc.

www.lorenzbooks.com

Publisher Joanna Lorenz
Managing Editor, Children's Books Gilly Cameron Cooper
Project Editor Rasha Elsaeed
Editorial Reader Joy Wotton
Introduction by Dr John Haywood
Authors Daud Ali, Charlotte Hurdman, Fiona Macdonald,
Lorna Oakes, Philip Steele, Michael Stotter, Richard Tames
Consultants Nick Allen, Clara Bezanilla, Felicity Cobbing,
Penny Dransart, Jenny Hall, Dr John Haywood, Dr
Robin Holgate, Michael Johnson, Lloyd Laing, Jessie Lim,
Heidi Potter, Louise Schofield, Leslie Webster,
Designers Simon Borrough, Matthew Cook, John Jamieson,
Joyce Mason, Caroline Reeves, Margaret Sadler, Alison Walker,
Stuart Watkinson at Ideas Into Print, Sarah Williams
Special Photography John Freeman
Stylists Konika Shakar, Thomasina Smith, Melanie Williams

Previously published as part of the *Step Into* series in 13 separate
volumes:
*Ancient Egypt, Ancient Greece, Ancient India, Ancient Japan,
Aztec & Maya Worlds, Celtic World, Chinese Empire, Inca World,
Mesopotamia, North American Indians, Roman Empire, The Stone
Age, Viking World*

PICTURE CREDITS
b=bottom, t=top, c=centre, l=left, r=right

AKG: 44tl, 45bl, 45br, 47tr, 57bl, 61l; The Ancient Art and
Architecture Collection Ltd: 11tr, 13tl, 22b, 28cl, 31bl, 33t,
42tl, 48cr, 57, 59tr; Ancient Egypt Picture library: 21t, 21br,
22t; Japan Archive: 43tl; The Bodleian Library; The Bridgeman
Art Library : 14cl, 14cr, 15tr, 39bl, 41tl, 41cl, 43cl, 56cl; Peter
Clayton: 11cr, 27tr, 29tl, 29cl; Bruce Coleman: 11bl, 35bl;
Corbis-Bettman: 60br; Corbis: 14tl, 15tl; C.M Dixon : 9tr, 9c,
10t, 10bl, 10br, 12l, 20l, 20r, 26cl, 26tr, 30l, 30r, 32cl, 32tr,
33bl, 42br, 44c, 45tl, 46tl, 60l, 61tr; Sue Cunningham
Photographic: 58; E.T Archive (Art Archive): 38bl, 40tr 49tr,
53tl, 53cl, 55c; Mary Evans Picture Library : 11tl, 27cl, 28tl,
28cr C.M Dixon: Werner Forman Archive: 35bl, 35br, 46bl,
47br, 52cr, 54l, 59bl; Geoscience Features Picture library: 31tr;
Michael Holford: 31tl, 32bl, 32br, 54cr; The Hutchison
Library: 43tl; Images of India: 14cl; Jenny Laing: 47bl;
Macquitty Collection: 38tr; Peter Newark's Pictures: 60tr;
Andes Press Agency: 56tr; Science Photo Library: 55cl; South
American Photo Library: 50c, 52c, 57tl, 57tr, 58t; Statens
Historik Museum: 49c; Still Pictures: 55tr, 56cr, 51tl; Zefa:
22l, 23t, 23b, 34tl, 35tr, 39tr, 39br

10 9 8 7 6 5 4 3 2 1

CONTENTS

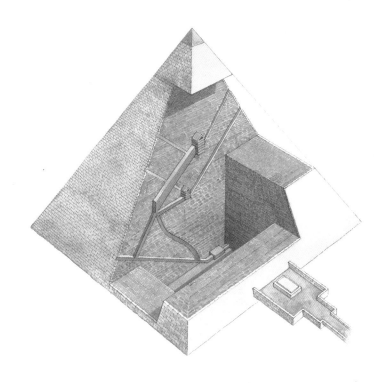

MAKING LIFE EASIER 4

The Beginning of Technology 8

Tools for the Job 10

Useful Crafts..............................12

Ceramic Skills in India 14

Scientific Minds in Mesopotamia 16

Mesopotamian Technology 18

Egyptian Calculations 20

The Pyramid Builders 22

Pyramid Construction 24

The Thinking Greeks 26

Greek Medical Foundations 28

Roman Empire Builders 30

Roman Healing Powers 32

Chinese Metalworkers 34

Chinese Firsts 36

Extraordinary Chinese Engineering 38

Chinese Science 40

Specialist Crafts in Japan 42

Fine Celtic Crafts 44

Celtic Metalworkers 46

Viking Crafts 48

Mesoamerican Time 50

Practical Incas 52

Inca Mining and Metalwork....54

Inca Medicine and Magic 56

Calculations Inca-style 58

Tribal Crafts in North America 60

GLOSSARY 62

INDEX 64

KEY

Look out for the border patterns used throughout this book. They will help you identify each culture.

The Stone Age	Ancient Rome	Aztec & Maya
India	Chinese Empire	Incas
Mesopotamia	Japan	North American
Ancient Egypt	Celts	Indians
Ancient Greece	Vikings	

Making Life Easier

Many of the animals of prehistoric times were faster, bigger, or stronger than the early humans who hunted them. The humans had one great advantage: they had bigger brains. The humans fought – and survived – by using their brains. They looked at the world they lived in and worked out how to make use of it. Their questioning, learning and understanding is what we call science. Their ability to shape and alter natural materials and turn them into tools, weapons and clothes was the beginning of technology.

The early species of human called *Homo habilis* (meaning handy man) was a pioneeer of technology. *Homo habilis* had big enough brains to work things out, and hands that could grip objects firmly. They were the first humans to make simple stone tools.

Early humans had to take meat they had hunted to the safety of a cave before other, fiercer animals came along. Smaller pieces were easier to carry.

Speed was vital. The humans found hard pebbles that they could split to reveal sharp edges to cut with. Technological breakthroughs such as this were gradually perfected and crafted into an ever-wider and more complex range of tools, all of which made human life safer and easier.

Because humans learned to make use of whatever local materials were available, they adapted to new environments instead of being confined to particular habitats like many animals. They moved all over the

The wheelbarrow was invented in China more than 1,000 years earlier than in the West. This was partly because there was a huge population of rice farmers who needed to find ways of making their back-breaking work easier.

Timeline 2,400,000BC–500BC

2,400,000 years ago. The early human *Homo habilis* (handy man) makes the first simple tools by splitting pebbles to create sharp edges.

stone spearheads

1,500,000 years ago. *Homo erectus* (upright man) makes hand axes, using flint, wood and bone.These are made over the next million years.

400,000 years ago. Date of the oldest surviving wooden tool - a spear from Germany.

100,000 years ago. Following the evolution of *Homo sapiens* – fully modern humans - a greater variety of more sophisticated tools are made.

round-based pot from ancient Japan

8000-3000BC Farming becomes widespread in Asia, Europe and Africa. New tools are invented such as axes to clear forests, sickles for harvesting wheat and grindstones for making flour. Pottery is used for cooking pots and storage containers.

6500BC The earliest known cloth is woven in Turkey.

6000BC The first metal tools and ornaments are made from copper in Turkey.

2,400,000BC   400,000BC   8000BC   6000B

world. The settlements and civilizations that grew up were often separated by oceans, mountains and deserts, but they often went through similar stages of discovery and invention quite independently of each other. The ancient Egyptians were using a wheel at around the same time as the Mesopotamians, for example. The Chinese, far in the East, invented a decimal system of mathematics from 300BC, although they had no known communication with the ancient Egyptians who had done the same thing about 350 years earlier.

The Chinese emperor Shenb Nong described 365 medicinal plants. The study of subjects such as medicine and astronomy was often left to the upper classes. Practical inventions were made by those who worked on the land or with their hands.

New technology often depended on the natural resources available. The earliest metalworkers were people who had noticed metal ore in the local rocks, and eventually learned how to extract it.

The speed with which science, craft and technology developed depended on what local materials were available, and how great the need for improvement was. Some Stone Age peoples became expert potters because they lived on clay deposits. At first, however, they only made little clay figures. It was many thousands of years before someone realised that clay could be made into ideal containers for food and liquid. The Chinese development of fine, waterproof pottery called porcelain 1,000 years before the West, was helped by rich supplies of a very fine clay called kaolin. Inca tribes became brilliant at working with gold and silver because of the rich resources of these precious metals in the South American mountains. Most new technology was invented by people who worked with their hands. Ancient China was one of the most inventive civilizations ever because it

3800BC The earliest bronze tools are made in Mesopotamia.

2650BC The first pyramid is built at Saqqara in Egypt.

stepped pyramid

1500BC The earliest iron tools are made in Turkey, and the earliest glass in Egypt and Mesopotamia.

1440BC Earliest metalworking (gold) in the Americas, in the Peruvian Andes.

750-54BC Celtic craftsmanship in gold and bronze flourishes in Europe.

Anglo-Saxon brooch

c. 700BC Babylonian astronomers identify the signs of the zodiac.

600BC The Chinese learn how to make cast iron.

c. 550BC Beginning of Greek science, mathematics and philosophy.

500BC Origins of the 260-day Mesoamerican calendar in Mexico.

Aztec pictures showing names for days

000BC   1500BC   600BC   500BC

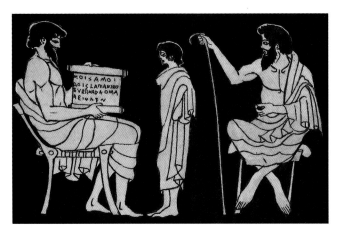

The ancient Greeks made learning a lot easier by inventing a simple alphabet. More people could read and write, so new ideas and technology spread. Ancient Greece had more scientists than any other early civilization.

was a huge country with great numbers of peasant farmers. The small, but very rich upper level of society demanded a luxurious life-style. This led to the development of fine craft skills, silk, and papermaking and printing for books.

In many civilizations, wealthy and powerful people were more interested in studying the stars and planets. Religion, astronomy and science were all closely connected in the early civilizations. The first astronomers in Babylon and Egypt believed that the movement of the stars and planets would help them to discover what the gods were planning. Greek mathematicians such as Pythagoras believed that they could understand more about the gods through numbers.

Greed and competition acted as spurs to science, craft and technology too. The person with the most efficient plow could till more land and harvest more crops than his neighbor. The country with the most deadly and effective weapons could build empires. Knowledge, science and craft skills developed very fast in empires that could call upon the resources of all the lands under their control. The rulers also had to keep in control, which was an incentive to keep one step ahead in road and transportation systems, and efficient ways of trading, language, writing, and coinage. And many

The Romans were great engineers, and built long, straight roads that were unsurpassed for centuries. A good road system was one way of keeping in control of their empire.

TIMELINE 500BC–AD1800

c. 400BC The Greek Hippocrates founds one of the earliest medical schools.

384-322BC Life of the Greek philosopher Aristotle. He is widely regarded as the founder of Western science.

Hippocrates, the Greek doctor

c. 310-230BC The Greek astronomer Aristarchus claims that the Earth goes around the Sun.

c. 200BC The Romans begin the large-scale use of concrete in building.

AD1-100 The magnetic compass is invented in China.

*c.*AD100 Paper is invented in China.

Chinese printed paper money

c. AD350 Maya astronomy develops in Mesoamerica.

AD605-9 The Grand Canal in China is completed.

AD800-1000 Viking craftsmanship in wood and precious metals flourishes in northern Europe.

carved prow of a Viking ship

500BC   200BC   AD300   AD800

kings and emperors, such as the Egyptian pharoahs, masterminded amazing engineering works so that they would be remembered for eternity.

Some cultures developed slowly because they did not have the way of life or the need to invent. Native peoples in North America did not invent sophisticated methods of transportation because they did not have much to carry. They did develop fine craft skills, though, making use of animal hide and bones, and dyes from plants.

As you read through these pages, you will be able to trace the special skills of different cultures, and discover what their particular contributions were to human knowledge. You will be able to see varying paces of technological development around the world and understand how different cultures had different priorities. Some enjoyed remarkably intensive periods of technological advancement that provided the foundations of much of the science, craft and technology that make modern life easier.

Although the Vikings were not great inventors, their dependence on the sea for travel and invasion meant that they developed very advanced boat-building skills.

New ideas and technology were spread by war. During the Crusade wars of the 11th and 12th centuries, peoples from northern Europe picked up tips on castle building and forging steel from the Muslim Saracen armies of the East.

AD810 The Persian mathematician al-Khwarizmi invents algebra.

AD850 Gunpowder is invented in China.

AD876 Indian mathematicians invent the symbol for zero

c. AD900-modern times. Pueblo Indian cultures in the Southwest of North America make fine painted pottery.

Inca gold llama

c. 1000 The Chinese invent moveable type printing.

c. 1150 The first mechanical clocks are made, in Europe.

c. 1200-1533 The Incas of the Andes make fine objects in gold, silver, textiles and pottery.

the astronomer Copernicus

1500-1800 The Scientific Revolution in Europe. Great advances in astronomy, physics, chemistry and biology are made.

chronometer

1700-1800 The Industrial Revolution begins the widespread use of steam power and machines .

1720 John Harrison develops the chronometer, an accurate timepiece.

1000   1500   1800

The Beginning of Technology

HANDY MAN
Chipped pebbles from Tanzania in Africa are some of the oldest tools ever found. They were made by *Homo habilis,* who lived almost two million years ago. *Homo habilis* (handy man) was the first human to make stone tools.

ABOUT TWO-AND-A-HALF MILLION years ago, early people began to chip stones to sharpen the edges to make tools. Flakes of hard stone were transformed into knives, spearheads, arrowheads, and other tools for everyday tasks. Hard, glassy stones such as flint made the toughest and sharpest tools. Flint was shaped and sharpened by knocking off flakes with a hammerstone. Hand-axes were used for digging and for chopping up animals. Smaller pieces of flint were used to scrape the flesh off animal skins. As people became more skillful, they made tools for chiseling and engraving. Tools like this made precise carving possible for other specialized tools such as harpoons, spear-throwers and needles made from bone.

FLAKING
Some 1½ million years after *Homo habilis,* a new species of human evolved with bigger brains. Neanderthals and *Homo sapiens* were far better toolmakers. They produced pointed or oval-shaped hand-axes (*left and middle*) and chopping tools (*right*).

VALUABLE STONE
Flint was dug in this mine in England from about 2800BC. Flint was vital for survival. People would trek long distances for the stone if there were none in their area.

MAKE A MODEL AX
You will need: self-drying clay, board, modeling tool, sandpaper, gray acrylic paint, wood stain, water pot, paintbrush, thick dowel, craft knife, ruler, chamois leather, scissors.

1 Pull out the clay into a thick block. With a modeling tool, shape the block into an ax head with a point at one end.

2 When the clay is completely dry, lightly rub down the ax head with sandpaper to remove any rough surfaces.

3 Paint the ax head a stone color, such as gray. You could use more than one shade if you like. Leave it to dry.

SPEAR POINT

Cro-Magnon people were top hunters, using leaf-shaped spear points to kill reindeer, wild horses, deer and woolly mammoths. They lived in Europe and Russia from around 38,000 years ago, and also learned to start fire by striking iron against flint.

STONES FOR TOOLS

The best rocks for tools were usually those that had been changed by heat. Obsidian, a glassy volcanic rock was widely used in the Near East and Mexico. It fractured easily, leaving sharp edges. In parts of Africa, quartz was made into beautiful, hardwearing hand-axes and choppers. Flint is another type of quartz. It is found in nodules in limestone rock, especially chalk. A hard igneous rock called diorite was used for making polished ax heads in Neolithic times.

quartz   *chert (a type of flint)*

AXES

Polished stone battle axes became the most important weapon in Scandinavia by the late Neolithic (New Stone Age) period. They date from about 1800BC.

TOOLMAKING LESSON

Stone Age people came to depend more and more on the quality of their tools. In this reconstruction, a father is passing on his skill in toolmaking to his son.

Prehistoric people used axes for chopping wood and cutting meat. They shaped a stone blade, then fitted it onto a wooden shaft.

4 Ask an adult to trim one end of a piece of thick dowel using a craft knife. Paint the piece with wood stain and leave to dry.

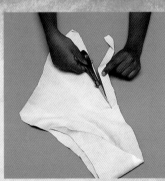

5 To bind the ax head to the wooden shaft, first carefully cut a long strip of leather about 1 in wide from a chamois cloth.

6 Place the ax head on the trimmed end of the shaft. Wrap the strip of leather around the head and shaft in a criss-cross pattern.

7 Pull the leather strip tight and wrap the ends twice around the shaft below the head. Tie the ends together and trim them.

Tools for the Job

DURING THE STONE AGE, wood, bone, antler and ivory were as important as stone for making tools and other implements. These softer materials were easily carved and shaped by stone tools. They could also be used to make more specialized stone tools. Bone and antler hammers and punches, for example, could achieve sharper cutting edges and more delicate flakes of stone. Flint blades were fitted into handles and mounts of wood. Antlers were converted into picks to dig up roots and lever lumps of rock from the ground. The broad shoulder-blade bones of cattle were turned into shovels, while smaller bones served as awls to punch small holes. Antlers and bones were also carved into spear-throwers. Ivory, from the tusks of the woolly mammoth, and bone could be crafted into fine-pointed needles, fish hooks, harpoon heads and knives. Adzes were tools for shaping wood and for making bows and arrows. Sometimes, things were made just for fun. Whistles and little paint holders were carved from small bones. Some pieces of stone, wood and bone were beautifully carved with pictures of the animals that were hunted and fine decorative patterns.

SPEAR-THROWER
This carving of a reindeer's head is probably part of a spear-thrower. The hunter slotted his spear into a hook at one end of the thrower and took aim. As he threw, the spear detached from the thrower and traveled farther and faster than if the hunter had thrown it just by hand. Hunting became safer and more accurate. Prehistoric carvers often incorporated the natural form of wood or bone into a design to suggest an animal's outlines.

SHAPING
An adze was a bit like an ax, except that its blade was at right angles to the handle. The flint blade on this adze dates from about 4000BC to 2000BC. Its wooden handle and binding are modern replacements for the originals, which have rotted away. Adzes were swung in an up-and-down movement and were used for jobs such as hollowing out tree trunks and shaping them to make dugout canoes.

AX
Early farmers needed axes to clear land for their crops. An experiment in Denmark using a 5,000-year-old ax showed that a man could clear 2½ acres of woodland in about five weeks. This ax head, dating from between 4000BC and 2000BC, has been given a modern wooden handle.

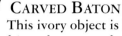

ANTLER PICK
Antlers were as useful to prehistoric humans as to the reindeer they came from! This tool comes from a Neolithic site near Avebury in England. Antler picks were used for digging and quarrying. Antler was a very hard material, but it could be carved into sharp spear points and barbed harpoons.

CRAFTSPEOPLE
A picture by an artist from the 1800s shows an imagined view of Stone Age. It shows tools being used and great care being taken over the work. Even everyday items were often finely carved and decorated by the craftspeople who made them.

CARVED BATON
This ivory object is known as a *bâton de commandement*. Several of these batons have been found, especially in France. But no one is sure what they were used for. Some experts think they were status symbols, showing the importance of the person carrying them. Others think that the holes were used to straighten arrows. Whatever their use, the batons are often decorated with fine animal carvings and geometric designs.

ANTLERS AT WORK
Two stags (male deer) fight. Male deer have large antlers, which they use to battle with each other to win territory and females. The stags shed and grow a new set of antlers each year, so prehistoric hunters and artists had a ready supply of material.

Useful Crafts

BASKET-MAKING WAS PROBABLY the very first handicraft. River reeds or twigs – whatever was found locally – were woven into shapes for carrying goods. Baskets were quick to make and easy to carry, but wore out easily and could not carry liquids. The discovery that clay turned into a hard, solid material when it was baked may have happened by accident, perhaps when a clay-lined basket was left in a bread oven. Although baked clay figures were made from about 24,000BC, it was thousands of years before pottery was used for cooking and storing food and drink. The first pots, shaped from coils or lumps of clay, were made in Japan around 10,500BC.

People learned how to spin thread from flax plants and animal hair. The loom, for weaving the thread into linen and wool cloth, was invented around 6000BC.

BAKED CLAY FIGURINE
This is one of the oldest fired-clay objects in the world. It is one of many similar figurines made around 24,000BC at Dolni Vestonice in the Czech Republic. Here, people hunted mammoths, woolly rhinoceroses and horses. They built homes with small, oval-shaped ovens, in which they fired their figurines.

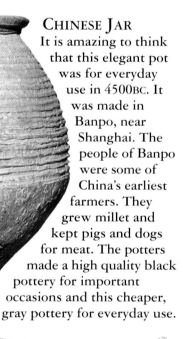

CHINESE JAR
It is amazing to think that this elegant pot was for everyday use in 4500BC. It was made in Banpo, near Shanghai. The people of Banpo were some of China's earliest farmers. They grew millet and kept pigs and dogs for meat. The potters made a high quality black pottery for important occasions and this cheaper, gray pottery for everyday use.

MAKE A CLAY POT
You will need: terra cotta modeling clay, wooden board, modeling tool, plastic flower pot, decorating tool, varnish, brush, sandpaper.

1 Roll out a long, thick sausage of clay on a wooden board. It should be at least ½ in in diameter.

2 Form the roll of clay into a coil to make the base of your pot. A fairly small base can be made into a pot, a larger one into a bowl.

3 Now make a fatter roll of clay . Carefully coil this around the base to make the sides of your pot.

HOUSEHOLD POTS

Many early pots were decorated with basket-like patterns. This one has a simple geometric design and was made in Thailand in about 3500BC. Clay pots like this were used for storing food, carrying water or cooking.

EASY TO CARVE

Steatite, or soapstone, has been used to make this carving from the Cycladic Islands of Greece. Soapstone is very soft and easy to carve. Figurines like this one were often used in funeral ceremonies. They could also be used either as the object of worship itself or as a ritual offering to a god. This figure has a cross around its neck. Although the symbol certainly has no Christian significance, no one really knows what it means.

WOVEN THREADS

The earliest woven objects may have looked like this rope and cane mat from Nazca in Peru. It was made around AD1000. Prehistoric people used plant-fiber rope to weave baskets and bags. The oldest known fabric dates from about 6500BC and was found at Çatal Hüyük in Turkey. Few woven objects have survived, as they rot quickly.

Fired-clay pots could only be made where there were natural deposits of clay. These areas seem to have specialized in baked-clay pottery and sculpture. The patterns used to decorate the pots vary from area to area.

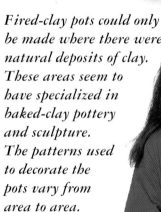

4 With a modeling tool, smooth down the edges of the coil to make it flat and smooth. Make sure there are no air spaces.

5 Place your pot over a flower pot to support it. Keep adding more rolls of clay to build up the sides of your pot.

6 Smooth down the sides as you add more rolls. Then use a decorating tool with a serrated end to make different patterns.

7 Leave your pot to dry out. When the clay is dry, varnish the outside. Use sandpaper to smooth the inside of your pot.

Ceramic Skills in India

CLAY AND TERRA COTTA OBJECTS (known as ceramics) play an important part in the study of history. Because they were fired (baked), traces of carbon (burned particles) are left on them. This enables archaeologists to date the objects quite accurately using a process called carbon-dating.

Making ceramics was one of the earliest crafts to be practised in India, for example. Many artefacts have been found at archaeological sites, dating as far back as 5000BC when there was a civilization in the Indus Valley. People made clay storage jars, terra cotta seals and terra cotta figurines of domestic animals. The animal figures may have been children's toys. Craftworkers also made terra cotta figurines of gods and goddesses, though they gradually began to make stone and metal images as well. Clay containers continued to be used throughout India's history. They kept food and liquids cool in the country's hot climate.

RECORD IN CLAY
This terra cotta cart found at the ancient city of Mohenjo-Daro is about 4,000 years old. Figurines like this have been found throughout the Indus Valley. They may have been toys, but they give clues to how people lived. For example, we can see that wheeled carts drawn by animals were in use.

IDEA EXCHANGE
Painted dishes like this one found on the plains of the River Ganges, date from between 1000 and 500BC. Similar pottery has been found across the region. This shows that people traveled and shared the same technology.

SIMILAR STYLES
Black and red painted pottery has been found at sites dating from the Indus Valley civilization, and at later sites dating from around 500BC. It is found all over the Indian subcontinent.

MAKE A WATER POT
You will need: inflated balloon, large bowl, strips of newspaper, flour and water or wallpaper paste, scissors, fine sandpaper, strip of corrugated cardboard, Scotch tape, terra cotta and black paint, paintbrushes, pencil, Elmer's glue.

1 Cover the balloon with 4 layers of newspaper soaked in paste. When dry, cut a slit in the papier-mâché. Remove the balloon. Add more layers to give the pot a tapered top.

2 Roll the corrugated cardboard into a circle shape to fit on to the narrow end of the pot to form a base. Fix the base in place with Scotch tape.

3 Cover the corrugated cardboard base with four layers of soaked newspaper. Leave to dry beween each layer. Smooth the edges with sandpaper.

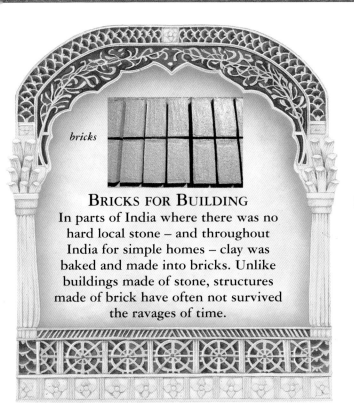

bricks

BRICKS FOR BUILDING

In parts of India where there was no hard local stone – and throughout India for simple homes – clay was baked and made into bricks. Unlike buildings made of stone, structures made of brick have often not survived the ravages of time.

TERRA COTTA GODDESS

A female terra cotta figure found in Mathura, Uttar Pradesh, may be an image of a mother goddess. Many terra cotta images were made during the Mauryan period (400–200BC) and immediately afterward. They were cheaper versions of the stone sculptures that were built at the imperial court.

THROWING A POT

A village potter shapes a clay vessel as it spins on his potter's wheel. Pottery was an important part of the ancient urban and village economies and is still practiced in India today. Clay used for making pottery is available in most parts of the land.

Clay water pots that are 4,000 years old have been found in the Indus Valley. People carried the pots on their heads.

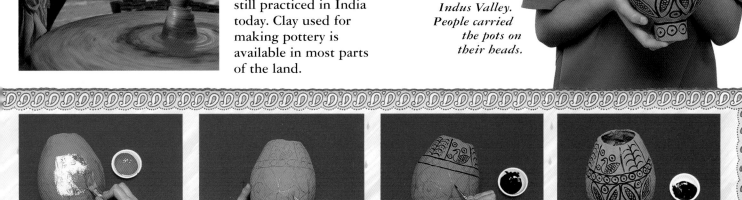

4 When it is dry, paint the water pot with two coats of terra cotta paint, to make it look as though it is made of terra cotta. Leave to dry between coats.

5 Draw some patterns on the water pot with a pencil. Copy the ancient Indian pattern shown here, or create your own individual design.

6 Carefully paint your designs using black paint and a fine paintbrush. Keep the edges of your lines neat and clean. Leave to dry.

7 Add final details, again using a fine paintbrush. When the paint is dry, seal the surface of the water pot with a coat of watered down Elmer's glue

Scientific Minds in Mesopotamia

THE SUMERIAN PEOPLE in Mesopotamia developed the world's first system of arithmetic around 2500BC. It was useful for making records of goods bought and sold. One number system used 10 as a base and the other, 60. They also calculated time in hour-long units of 60 minutes. Sumerian astronomers worked out a calendar based on 12- and 28-day cycles and 7-day weeks from studying the moon and the seasons. Later, the Babylonians made a detailed study of the heavens, and could predict events such as eclipses.

Mesopotamian doctors did not fully understand how the body worked, but they did make lists of symptoms. Their observations passed on to the Greeks centuries later and so became one of the foundations of modern medicine.

HEAVY COUGH CURE
Inscriptions on a clay tablet suggest mixing balsam (a herb) with strong beer, honey and oil to cure a cough. The mixture was taken hot, without food. Then the patient's throat was tickled with a feather to make him sick. Other prescriptions used mice, dogs' tails and urine.

BAD OMEN
Mesopotamians thought that eclipses were a bad sign – unless they were obscured by cloud. If an eclipse was covered by cloud in a particular city, the local king was told that it had nothing to do with him or his country.

MEDICINAL BREW
Servants are distilling essence of cedar, a vital ingredient for a Mesopotamian headache cure. Cedar twigs were heated to give off a vapor. This condensed against the cooler lid and trickled into the rim of the pot from where it was collected. The essence was mixed with honey, resin from pine, myrrh and spruce trees, and fat from a sheep's kidney.

MAKE A SET OF LION WEIGHTS
You will need: pebbles of various sizes, kitchen scales, modeling clay, cutting board, cocktail stick, paints and paintbrushes.

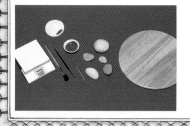

1 Weigh a pebble and add modeling clay to make it up to a weight of 8 oz. Once the clay has dried out, the final weight will be only about 7 oz.

2 Take a portion of the weighed modeling clay and shape it into a rectangle roughly 4½ in by 2¾ in. This will be the base for your weight.

3 Wrap another piece of the weighed modeling clay around the weighed pebble to make the lion's body. Shape the body into a pear shape.

SKY MAP

The sky in this Mesopotamian astronomical map is divided into eight parts and the stars in each section are indicated. The heavens were seen as a source of information about the future, so the kings often consulted astronomers. One astronomer wrote to the king in the 600s BC: "I am always looking at the sky but nothing unusual has appeared above the horizon."

MEDICINE MEN

Mesopotamian doctors gave their patients a bultu to cure them. A bultu was a mushy mixture of herbs and other things including perhaps beetroot, coriander and parsley. It could be drunk or used externally. One medical text lists 230 different drugs and herbs. Unfortunately most of them cannot be translated.

beet   *coriander*

parsley

WEIGHTS AND MEASURES

Officials weigh metal objects that have been taken as booty after a victory. The duck-shaped object is a weight. The kings were responsible for seeing that weights and measures were exact and that nobody cheated customers. Prices were fixed by law and calculated in shekels (1 shekel was about 3/10 oz of silver).

Bronze lion weights from a set belonging to King Shalmaneser V have been found at Nimrud. Like your weights they were of different sizes.

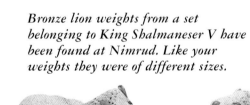

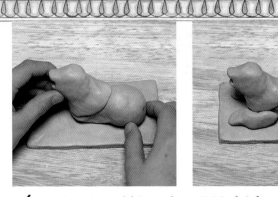

4 Position the pebble and clay on to its base. Add another piece of weighed clay to form the head and mane. Shape the face and jaw with your fingers.

5 Model four pieces of weighed clay to make the lion's four legs and stick them on to the body. Flatten the clay slightly at each end for the paws.

6 Make a tail and ears using up the remaining weighed clay. Using the cocktail stick, add extra detail to the face, mane, paws and tail. Leave to dry.

7 Paint the lion and the base cream. Flick with brown paint for a mottled appearance. Add details to the face, mane and paws. Make more lions for a set.

SCIENTIFIC MINDS IN MESOPOTAMIA **17**

Mesopotamian Technology

A WHOLE CULTURE OF SPECIALIST craftworkers grew up in Mesopotamia and quickened the pace of improvement and invention. The Sumerians learned to make pottery by shaping the wet clay on a potter's wheel by about 3500BC, and became experts at making cloth, leatherwork and making fine jewelry. They were among the first people in the world to use metal. A copper sculpture of a lion-headed eagle found near the ancient city of Ur, dates from around 2600BC. Mesopotamian armies used weapons and armor of bronze, an alloy of copper and tin, which is stronger than plain copper. The Mesopotamians were also experts at irrigation and flood control, building elaborate canals, and water storage and drainage systems.

SUPPLYING THE CITY

Water wheels and aqueducts like these are still used in the Middle East today. The Assyrians built aqueducts to take water to the cities to meet the needs of their growing populations. The Assyrian king Sennacherib had 6 miles of canals cut. They led from the mountains to the city of Nineveh. He built dams and weirs to control the flow of water, and created an artificial marsh, where he bred wild animals and birds.

A WEIGHTY CHALLENGE

Workers in a quarry near the Assyrian city of Nineveh prepare to move an enormous block of stone roughly hewn in the shape of a lamassu (human-headed winged bull). The stone is on a sledge carried on wooden rollers. At the back of the sledge, some men have thrown ropes over a giant lever and pull hard. This raises the end of the sledge and other workers push a wedge underneath. More workers stand ready to haul on ropes at the front of the sledge. At a signal everyone pulls or pushes and the sledge moves forward.

MAKE A PAINTED PLATE

You will need: a plate, flour, water and newspaper to make papier-mâché, scissors, pencil, fine sandpaper, ruler, paints and paintbrushes.

1 Tear strips of newspaper and dip them in the water. Cover the whole surface of the plate with the wet newspaper strips.

2 Mix up a paste of flour and water. Cover the newspaper strips with the paste. Allow to dry, then add two more layers, leaving it to dry each time.

3 When the papier-mâché is dry, trim around the plate to make a neat edge. Remove the plate. Add more papier-mâché to strengthen the plate.

MAKING CLOTH

Spinning and weaving were usually done by women in the home or in state or temple factories. Large herds of sheep and goats were kept to produce wool, to make clothing. Flax was grown for its fibers, which were used to make linen as early as 3000BC. Cotton was not introduced until the reign of the Assyrian king, Sennacherib, in the 700s BC.

sheep's wool

sheep's fleece

wool

METALWORKERS

Ceremonial daggers demonstrate the Sumerians' skill at working with gold as far back as 2600BC. Real weapons had bronze blades. The Sumerians made a wax model of the object required. They covered this with clay to make a mold. They heated the mold to harden the clay. The melted wax was poured out through a small hole, and molten metal poured in to replace it. When cool, the clay mold was broken, to reveal the metal object inside.

You have copied a plate from Tell Halaf, a small town where some of the finest pots in the ancient world were made. They were decorated with orange and brown paints made from oxides found in clay.

HAND-MADE VASES

Vases found in Samarra in the north of Mesopotamia were produced about 6,000 years ago. They were shaped by hand and fired in a kiln, then painted with geometric designs. Later, a wheel like a turntable was used to shape the clay, which speeded up the process.

4 When the papier-mâché is completely dry, smooth it down with fine sandpaper. Then paint the plate on both sides with a white base coat.

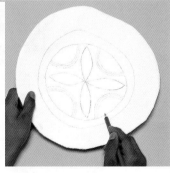

5 When the paint is dry, use a pencil and ruler to mark a dot in the center of the plate. Draw four large petals around this point and add details as shown above.

6 When you are happy with your design, paint in the patterns using three colors for the basic pattern. Allow each color to dry before adding the next.

7 Add more detail to your plate, using more colors, including wavy lines around the edge. When you have finished painting, leave it to dry.

Egyptian Calculations

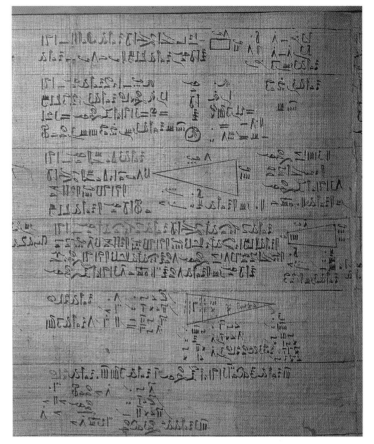

MATHEMATICAL PAPYRUS
This papyrus shows methods for working out the areas of squares, circles and triangles. It dates from around 850BC. These methods would have been used in calculations for land areas and pyramid heights on Egyptian building projects. Other surviving writings show mathematical calculations for working out how much grain might fit into a store. The Egyptians used a decimal system of numbering with separate symbols for one, ten, 100 and 1,000. Eight was shown by eight one symbols – 11111111.

T HE ANCIENT EGYPTIANS had advanced systems of numbering and measuring. They put this knowledge to good use in building, engineering and surveying the land. However, their knowledge of science was often mixed up with superstitions and belief in magic. For example, doctors understood a lot about broken bones and surgery, but at the same time they used all kinds of spells, amulets (charms) and magic potions to ward off disease. Much of their knowledge about the human body came from their experience of preparing the dead for burial.

The priests studied the stars carefully. They thought that the planets must be gods. The Egyptians also worked out a calendar, which was very important for working out when the Nile floods were due and when to plant crops.

CUBIT MEASURE
Units of measurement included the royal cubit of about 21 in and the short cubit of 18 in. A cubit was the length of a man's forearm and was subdivided into palms and fingers.

MAKE A WATER CLOCK

You will need: self-drying clay, plastic flowerpot, modeling tool, skewer, pencil, ruler, masking tape, scissors, yellow acrylic paint, varnish, water pot and brush. Optional: rolling pin and board.

1 Begin by rolling out the clay. Take the plastic flowerpot and press its base firmly into the clay. This will be the bottom of your water clock.

2 Cut out an oblong of clay large enough to mold around the flowerpot. Add the base and use your modeling tool to make the joints smooth.

3 Make a small hole near the bottom of the pot with a skewer, as shown. Leave the pot in a warm place to dry. When the clay has dried, remove the flowerpot.

How Deep is the River?

A series of steps called a nilometer was used to measure the depth of water in the River Nile. The annual floods were desperately important for the farmers living alongside the Nile. A good flood measured about 23 ft. More than this and farm buildings and channels might be destroyed. Less, and the fields might go dry.

Star of the Nile

This astronomical painting is from the ceiling of the tomb of Seti I. The study of the stars, as in most of the ancient civilizations, was part religion, part science. The brightest star in the sky was Sirius, which we call the dog star. The Egyptians called it Sopdet, after a goddess. This star rose into view at the time when the Nile floods were due and was greeted with a special festival.

Medicine

Most Egyptian medicines were based on plants. One cure for headaches included juniper berries, coriander, wormwood and honey. The mixture was rubbed into the scalp. Other remedies included natron (a kind of salt), myrrh and even crocodile droppings. Some Egyptian medicines probably did heal the patients, but others did more harm than good.

coriander

garlic

4 Mark out lines at ⅛ in intervals inside the pot. Mask the ends with tape and paint the lines yellow. When dry, remove the tape. Ask an adult to varnish the pot inside.

5 Find or make another two pots and position them as shown. Ask a partner to put their finger over the hole in the clock while you pour water into it.

6 Now ask your partner to take their finger away. The length of time it takes for the level of the water to drop from mark to mark is the measure of time.

Time was calculated on water clocks by calculating how long it took for water to drop from level to level. The water level lowered as it dripped through the hole in the bottom of the pot.

The Pyramid Builders

LITTLE IN THE ANCIENT WORLD matched the Egyptian pyramids for size and engineering achievement. They were massive four-sided tombs, built for the pharaohs of the Old Kingdom (2686-2181BC). Each side, shaped like a triangle, met together in a point at the top to make the pyramid shape. The most impressive pyramids, built at Giza near the modern city of Cairo, had flat sides. Their summits were probably capped in gold. Inside the pyramids were burial chambers and secret passages. The Egyptians may have seen the pyramid shape as a stairway to heaven to help the pharaoh on his journey to the afterlife.

The pyramids were built with fantastic skill and mathematical accuracy by teams of architects, engineers and stonemasons. They still stand today. The manual labor was provided not by slaves, but by about 100,000 ordinary people. These unskilled workers had to offer their services each year when the flooding Nile made work in the fields impossible.

WORN DOWN BY THE WIND
This pyramid at Dahshur was built for pharaoh Amenemhat III. Once the limestone casing had been stolen, its mud-brick core was easily worn down by the harsh desert winds. Pyramids had become popular burial monuments after the building of the first step pyramid at Saqqara. Examples can be seen at Maidum, Dahshur and Giza. However, Amenemhat's pyramid is typical of those built during the Middle Kingdom when inferior materials were used.

THE STEP PYRAMID
The earliest step pyramid was built at Saqqara for the pharaoh Zoser. The tomb probably started out as a mastaba, an older type of burial site made up of a brick structure over an underground tomb. The upper levels of Zoser's mastaba were redesigned as a pyramid with six huge steps. It was 60m high and towered above the desert sands. It covered the underground tomb of the pharaoh and included 11 burial chambers for the other members of the royal family.

ROYAL ARCHITECT
Imhotep was vizier, or treasurer, in the court of the great pharaoh Zoser. He designed the huge step pyramid at Saqqara. This pyramid was one of the first large monuments made entirely of stone. Imhotep was also a wise man who was an accomplished scribe, astronomer, doctor, priest and architect. He was worshipped as a god of medicine. and architecture.

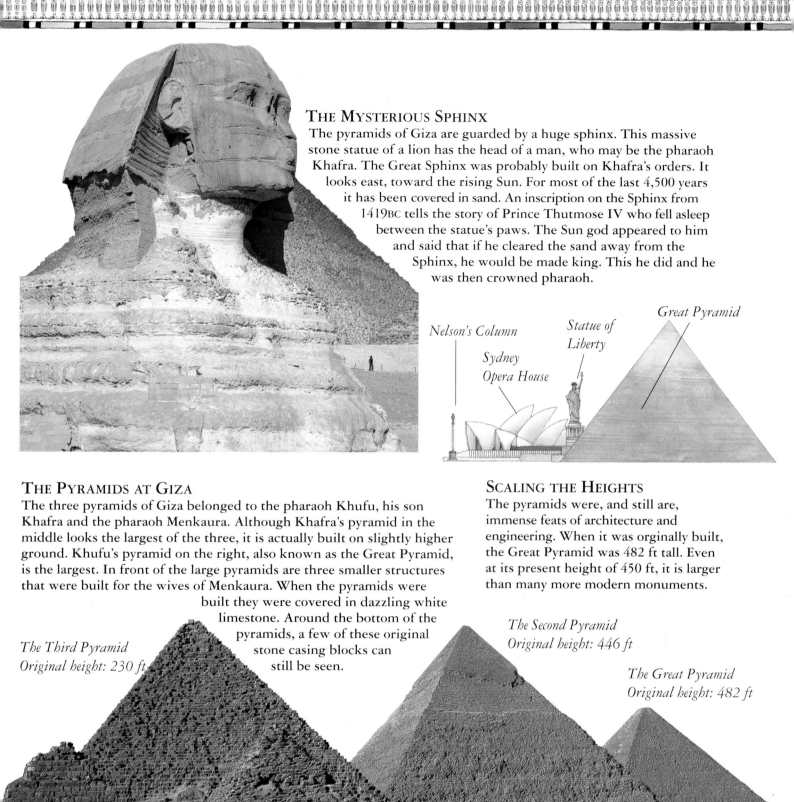

THE MYSTERIOUS SPHINX

The pyramids of Giza are guarded by a huge sphinx. This massive stone statue of a lion has the head of a man, who may be the pharaoh Khafra. The Great Sphinx was probably built on Khafra's orders. It looks east, toward the rising Sun. For most of the last 4,500 years it has been covered in sand. An inscription on the Sphinx from 1419BC tells the story of Prince Thutmose IV who fell asleep between the statue's paws. The Sun god appeared to him and said that if he cleared the sand away from the Sphinx, he would be made king. This he did and he was then crowned pharaoh.

Nelson's Column

Sydney Opera House

Statue of Liberty

Great Pyramid

THE PYRAMIDS AT GIZA

The three pyramids of Giza belonged to the pharaoh Khufu, his son Khafra and the pharaoh Menkaura. Although Khafra's pyramid in the middle looks the largest of the three, it is actually built on slightly higher ground. Khufu's pyramid on the right, also known as the Great Pyramid, is the largest. In front of the large pyramids are three smaller structures that were built for the wives of Menkaura. When the pyramids were built they were covered in dazzling white limestone. Around the bottom of the pyramids, a few of these original stone casing blocks can still be seen.

SCALING THE HEIGHTS

The pyramids were, and still are, immense feats of architecture and engineering. When it was orginally built, the Great Pyramid was 482 ft tall. Even at its present height of 450 ft, it is larger than many more modern monuments.

The Third Pyramid
Original height: 230 ft

The Second Pyramid
Original height: 446 ft

The Great Pyramid
Original height: 482 ft

Pyramid Construction

QUARRYING

The core of the pyramid was of rough stone, from local quarries. Better quality stone was shipped from Aswan, 600 mi away. Workers used wooden mallets to drive wedges and chisels into the stone to split it.

FOR MANY YEARS the Great Pyramid at Giza was the largest building in the world. Its base is about 276 sq yd, and its original point was 582 ft high. It is made up of about 2,300,000 massive blocks of stone, each one weighing nearly 3 tons. The blocks were secured by rope on wooden rollers. Laborers hauled them up ramps of solid earth that were built up the side of the pyramid as it grew higher and higher. The ramps were destroyed when the pyramid was completed.

The Great Pyramid was amazingly accurate and symmetrical in shape. The land had to be absolutely flat for this to be possible. The Egyptians cut channels across the building site and filled them with water. They used the water line as a marker for making the site level. The four corners of the pyramid are aligned exactly to face north, south, east and west. This was worked out by astronomers, by observing the stars.

INSIDE WORK

Inside the pyramid were narrow passages, and tomb chambers. They were lined with good quality granite. Flickering light came from pottery lamps that consisted of a wick of twine or grass soaked in animal or fish oil. Bundles of papyrus (reeds) were dipped in resin or pitch and lit to use as torches.

MAKE A PYRAMID

You will need: card, pencil, ruler, scissors, Elmer's glue and brush, masking tape, acrylic paint (yellow, white, gold), plaster paste, sandpaper, water pot and brush.

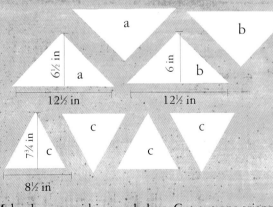

a   b

6½ in   a   6 in   b

12½ in   12½ in

c   c

7¾ in   c   c

8½ in

Make the pyramid in two halves. Cut out one triangle (a) for the base, one triangle (b) for the inside and two of triangle (c) for the sides of each half section.

1 Glue the half section of the pyramid together, binding the joints with pieces of masking tape, as shown. Now make the second half section in the same way.

INSIDE A PYRAMID

This cross-section shows the inside of the Great Pyramid. The design of the interior changed several times during its construction. An underground chamber may originally have been intended as the pharaoh Khufu's burial place. This was never finished. The Queen's Chamber was also found empty. The pharaoh was actually buried in the King's Chamber. Once the funeral was over, the tomb was sealed from the inside to prevent people breaking in. Blocks of stone were slid down the Grand Gallery. The workmen left through a shaft and along a corridor before the stones thudded into place.

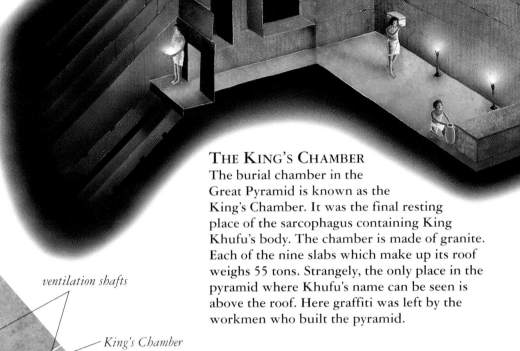

THE KING'S CHAMBER

The burial chamber in the Great Pyramid is known as the King's Chamber. It was the final resting place of the sarcophagus containing King Khufu's body. The chamber is made of granite. Each of the nine slabs which make up its roof weighs 55 tons. Strangely, the only place in the pyramid where Khufu's name can be seen is above the roof. Here graffiti was left by the workmen who built the pyramid.

ventilation shafts

King's Chamber

Grand Gallery

Queen's Chamber

escape shaft for workers

corridor

unfinished chamber

2 Mix up yellow and white paint with a little plaster paste to achieve a sandy texture. Then add a little glue so that it sticks to the card. Paint the pyramid sections.

3 Leave the painted pyramid sections to dry in a warm place. When they are completely dry, sand down the tips until they are smooth and mask them off with tape.

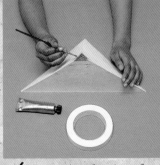

4 Now paint the tips of each half of the pyramid gold and leave to dry. Finally, glue the two halves together and place your pyramid on a bed of sand to display.

The building of the Great Pyramid probably took about 23 years. Originally the pyramids were cased in pale limestone, so they would have looked a brilliant white. The capstone at the very top of the pyramid was probably covered in gold.

The Thinking Greeks

THE ANCIENT GREEKS COULD AFFORD time for studying and thinking because their civilization was both wealthy and secure. They learned astrology from the Babylonians and mathematics from the Egyptians. They used their scientific knowledge to develop many practical inventions, including water clocks, cogwheels, gearing systems, slot machines and steam engines. However, these devices were not widely used as there were many slave workers to do the jobs.

The word "philosophy" comes from the Greek word *philosophos*, meaning love of knowledge. The Greeks developed many different branches of philosophy. Three of these were politics (how best to govern), ethics (how to behave well) and cosmology (how the universe worked).

Greek philosophers recognized the value of experimenting. But they could not always see their limitations. Aristotle discovered that evaporation turned salt water into fresh water, and wrongly assumed wine would turn into water by the same process.

GREAT THINKER
The philosopher Aristotle (384–322BC) is often recognized as the founder of Western science. He pioneered a rational approach to the world, that was based on observing and recording evidence. For three years, he was the tutor of Alexander the Great.

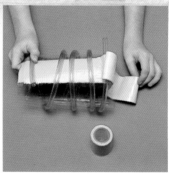

CLOCK TOWER
The Tower of the Winds in Athens contains a water clock. The original Egyptian invention was a bucket of water with a tiny hole in the bottom. As the water dripped out of it, the water level fell past scored marks on the inside of the bucket, measuring time. The Greeks improved on this design, using the flow of water to work a dial with a moving pointer.

ARCHIMEDES SCREW
You will need: clean, empty plastic bottle, scissors, modeling clay, strong tape, length of clear plastic tube, bowl of water, blue food coloring, empty bowl.

1 Cut off the bottle top. Place the modeling clay into the middle of the bottle, about 2 in from the end. Punch a hole here with the scissors.

2 Cut a strip of tape the same length as the bottle. Tape it to the middle of the bottle. This will give the tube extra grip later on.

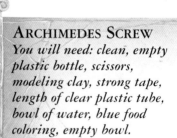

3 Twist the length of tube around the bottle. Go from one end of the bottle to the other. Tape the tube into place over the first piece of tape.

WATER LIFTER

When an Archimedes screw is turned, it lifts water from one level to another. It is named after its inventor, the scientist Archimedes, who lived about 287–211BC, in Syracuse, Sicily. The device is still used today.

FATHER OF GEOMETRY

Euclid (about 330–260BC) was a mathematician. He lived in the Greek-Egyptian city of Alexandria. He is known as the father of geometry, which comes from the Greek word for "measuring land." Geometry is the study of points, lines, curves, surfaces and their measurements. His geometry textbook was called *Elements*. It was still widely used in the early part of the 1900s, over 2,000 years after Euclid's death. This picture shows the front page of an edition of the book that was printed in London in 1732.

4 Place a few drops of the blue food coloring into the bowl of water. Stir it in so that the color mixes evenly throughout the water.

5 Place one end of the bottle into the bowl of blue water. Make sure that the tube at the opposite end is pointing toward the empty bowl.

6 Twist the bottle around in the blue water. As you do so, you will see the water start traveling up the tube and gradually filling the other bowl.

The invention of the Archimedes screw made it possible for farmers to water their fields with irrigation channels. It saved them from walking back and forth to the river with buckets.

Greek Medical Foundations

THE ANCIENT GREEKS LAID the foundations of modern medicine. Although they believed that only the gods had the power to heal wounds and cure sickness, they also developed a scientific approach to medicine. Greek doctors treated injuries and battle wounds by bandaging and bone-setting. They prescribed rest, diet and herbal drugs to cure diseases, although they were powerless against epidemics, such as plague. Doctors believed that good health was dependent on the balance between four main body fluids – blood, phlegm and yellow and black bile. If this balance was disturbed, they attempted to restore it by applying heated metal cups to the body to draw off harmful fluids. This mistaken practice continued in Europe until the 1600s.

FATHER OF MEDICINE
Hippocrates founded a medical school around 400BC. He taught that observation of symptoms was more important than theory. His students took an oath to use their skills to heal and never to harm. Doctors still take the Hippocratic oath today.

BODY BALANCE
Bleeding was a common procedure, intended to restore the body's internal balance. This carving shows surgical instruments and cups used for catching blood. Sometimes bleeding may have helped to drain off poisons, but more often it can only have weakened the patient.

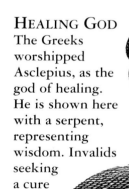

HEALING GOD
The Greeks worshipped Asclepius, as the god of healing. He is shown here with a serpent, representing wisdom. Invalids seeking a cure made a visit to his shrine.

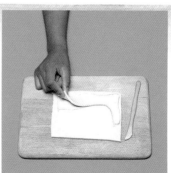

LEG OFFERING
You will need: self-drying modeling clay, rolling pin, board, ruler, modeling tool, paintbrush, cream acrylic paint.

1 Divide the clay into two pieces. With the rolling pin, roll out one piece to 6 in length, 4 in width and ¾ in depth. This is the base for the leg.

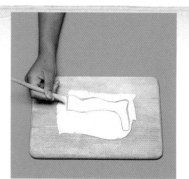

2 Roll out the second piece of clay. With the modeling tool, carve out a leg and foot shape. It should be big enough to fit on one side of the base.

3 Gently place the leg on the right-hand side of the base. With the tool, draw a shallow outline around the leg into the base. Remove the leg.

THEORY AND PRACTICE

Patients would explain their dreams to doctors, who then prescribed treatment. In this relief, a healing spirit in the shape of a serpent visits a sleeping patient. In the foreground, the physician bandages the wounded arm.

TOOL KIT

The Greeks used bronze surgical instruments, including forceps and probes. Surgery was usually a last resort. Even when it was successful, patients often died from the shock and pain, or from infection afterward. Operations on limbs were more successful than those on body cavities such as the chest or stomach.

NATURAL HEALING

The Greeks used a large variety of natural treatments to cure illnesses. Herbal remedies were particularly popular. Lentils, mustard and honey may have been combined in a poultice and applied to a wound.

lentils

honey

mustard

This model is based on a real one that was left as a thanks offering to the god Asclepius by someone whose leg was affected by illness. This was a common practice in ancient Greece.

4 With the tool, score the outline with lines. Carve the ancient Greek message seen in the picture above next to the leg.

5 Mold the leg onto the scored area of the base. Use your fingers to press the sides of the leg in place. Carve toes and toenails into the foot.

6 Paint over the entire leg offering with a cream color, to give it an aged look. Leave to dry overnight. Your leg offering is done.

Roman Empire Builders

THE ROMANS ADOPTED many of the ideas of the Greeks, such as the principles of architecture, and developed them further. They built magnificent domes, arched bridges and grand public buildings throughout the Empire, spreading their ideas and their expert skills.

They built long, straight roads to carry supplies, and messengers to the farthest corners of the Empire. The roads had a slight hump in the middle so that rainwater drained to the sides. Some were paved with stone and others were covered with gravel or stone chippings. Engineers designed aqueducts to carry water supplies to their cities. If possible, local stone and timber were used for building works. The Romans were the first to develop concrete, which was cheaper and stronger than stone. The rule of the Romans came to an end in western Europe over 1,500 years ago. Yet many of their techniques and principles of building are still in use today.

ROMAN ROADS
A typical Roman road stretches into the distance. It runs through the town of Ostia, in Italy. Roman road-building techniques remained unmatched in Europe until the 1800s.

MUSCLE POWER
Romans used big wooden cranes to lift heavy building materials. The crane is powered by a huge treadwheel. Slaves walk around and around in the wheel, making it turn. The turning wheel pulls on the rope, that is tied around the heavy block of stone, raising it off the ground.

MAKE A GROMA

You will need: large, strong piece of cardboard, scissors, ruler, pencil, square of card, Elmer's glue, masking tape, balsa wood pole, Plasticine, silver foil, string, large sewing needle, acrylic paints, paintbrush, water pot, broom handle.

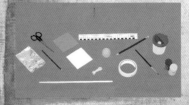

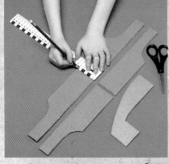

1 Cut out three pieces of cardboard – two 8 in x 2½ in, one 16 in x 2½ in. Cut another piece, 6 in x 4½ in, for the handle. Then cut them into shape, as shown above.

2 Measure to the center of the long piece. Use a pencil to make a slot here, between the layers of cardboard. The slot is for the balsa wood pole.

3 Slide the balsa wood pole into the slot and tape the cardboard pieces in a cross. Use the card square to make sure the four arms of the groma are at right angles. Glue in place.

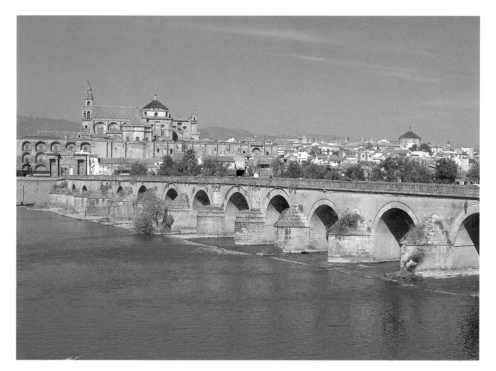

BUILDING MATERIALS

The Romans used a variety of stones for building, usually from local quarries. Limestone and a volcanic rock called tufa were used in the city of Pompeii. Slate was used for roofing in parts of Britain. Fine marble, used for temples and other public buildings, was available in the Carrara region of Italy, as it still is today. Marble was also imported from overseas.

marble

slate

ARCHING STRENGTH

The Roman bridge over the River Guadalquivir at Cordoba in Spain still stands today. The arch was a key element in many Roman buildings, including domed roofs. It gave stronger support than a simple beam.

WALLS OF ROME

The city of Rome's defences were built at many stages in its history. These sturdy walls were raised during the reign of the Emperor Marcus Aurelius, AD121–180. The Aurelian Walls were so well built that they are still in good condition.

Slot the arms on to the balsa wood pole. Use the plumb lines as a guide to make sure the pole is vertical. The arms can then be used to line up objects in the distance. Romans used a groma to measure right angles and to make sure roads were straight.

4 Roll the Plasticine into four small cones and cover them with foil. Thread string through the tops, as shown. These are the groma's plumb lines, or vertical guides.

5 Tie the plumb lines to each arm, as shown. They must all hang at the same length – 8 in will do. If the Plasticine is too heavy, use wet newspaper rolled up in the foil.

6 Split the top of the handle piece, and wrap it around the balsa wood pole. Glue it in place, as shown. Do the same on the other end with the broom handle. Paint the groma.

Roman Healing Powers

SOME ROMANS lived to old age, but most died before they reached the age of 50. Archaeologists have found out a lot about health and disease in Roman times by examining skeletons that have survived. They can tell, for example, how old a person was when he or she died and their general state of health during life. Ancient writings also provide information about Roman medical knowledge. Roman doctors knew very little science. They healed the sick through a mixture of common sense, trust in the gods and magic. Most cures and treatments had come to Rome from the doctors of ancient Greece. The Greeks and Romans also shared the same god of healing, Aesculapius (the Greek Asclepius). There were doctors in most parts of the Empire, as well as midwives, dentists and eye specialists. Surgeons operated on wounds received in battle, on broken bones and even skulls. The only pain killers were made from poppy juice.

GODDESS OF HEALTH
Greeks and Romans honored the daughter of the god Aesculapius as a goddess of health. She was called Hygieia. The word hygienic, which comes from her name, is still used today to mean free of germs.

A PHARMACY
This pharmacy is run by a woman. This was quite unusual for Roman times, as women were rarely given positions of responsibility. Roman pharmacists collected herbs and often mixed them for doctors.

MEDICINE BOX
Boxes like this one would have been used by Roman doctors to store various drugs. Many of the treatments used by doctors were herbal, and not always pleasant to take!

MEDICAL INSTRUMENTS
The Romans used a variety of surgical and other instruments. These are made in bronze and include a scalpel, forceps and a spatula for mixing and applying various ointments.

TAKING THE CURE

These are the ruins of a medical clinic in Asia Minor (present-day Turkey). It was built around AD150, in honor of Aesculapius, the god of healing. Clinics like this one were known as therapy buildings. People would come to them seeking cures for all kinds of ailments.

BATHING THE BABY

This stone carving from Rome shows a newborn baby being bathed. The Romans were well aware of the importance of regular bathing in clean water. However, childbirth itself was dangerous for both mother and baby. Despite the dangers, the Romans liked to have large families, and many women died giving birth.

HERBAL MEDICINE

Doctors and traveling healers sold all kinds of potions and ointments. Many were made from herbs such as rosemary, sage and fennel. Other natural remedies included garlic, mustard and cabbage. Many of the remedies would have done little good, but some of them did have the power to heal.

garlic

sage

rosemary

Chinese Metalworkers

THE CHINESE MASTERED THE secrets of making alloys (mixtures of two or more metals) during the Shang Dynasty (*c*.1600BC–1122BC). They made bronze by melting copper and tin to separate each metal from its ore, a process called smelting. Nine parts of copper were then mixed with one part of tin and heated in a charcoal furnace. When the metals melted, they were piped into clay molds. Bronze was used to make objects such as ceremonial pots, statues, bells, mirrors, tools and weapons.

By about 600BC, the Chinese were smelting iron ore. They then became the first people to make cast iron – around 1500 years before the process was discovered in the West – by adding carbon to the molten metal. Cast iron is tougher than bronze and was soon being used to make weapons, tools and plow blades. By AD1000, the Chinese were mining and working a vast amount of iron. Coke (a type of coal) had replaced the charcoal used in furnaces, which were fired up by water-driven bellows.

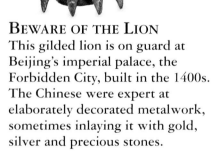

BEWARE OF THE LION
This gilded lion is on guard at Beijing's imperial palace, the Forbidden City, built in the 1400s. The Chinese were expert at elaborately decorated metalwork, sometimes inlaying it with gold, silver and precious stones.

SILVER SCISSORS
This pair of scissors is made of silver. They are proof of the foreign influences that entered China in the AD700s, during the boom years of the Tang dynasty. The metal is beaten, rather than cast in the Chinese way. It is decorated in the Persian style of the Silk Road, with engraving and punching.

MAKE A NECKLACE

You will need: tape measure, thick wire, thin wire, masking tape, scissors, tinfoil, measuring spoon, glue and brush, thin wire.

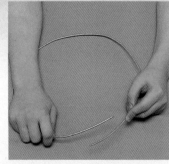

1 Measure around your neck using a tape measure. Ask an adult to cut a piece of thick wire to 1½ times this length. Shape it into a rough circle.

2 Cut two 1½ in pieces of thin wire. Coil loosely around sides of thick wire. Tape ends to thick wire. Slide thick wire through coils to adjust fit.

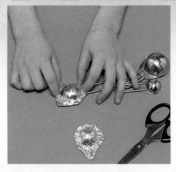

3 Cut out an oval-shaped piece of tinfoil. Shape it into a pendant half, using a measuring spoon or teaspoon. Make 9 more halves.

MINERAL WEALTH
The Chinese probably learned to smelt ore in furnaces from their experience with high-temperature pottery kilns. The land was rich in copper, tin and iron, and the Chinese were very skilled miners. Large amounts of precious metals, such as gold and silver, had to be imported.

gold nugget   *silver ore*

DECORATIVE PROTECTION
A network of gold threads makes up these fingernail protectors of the 1800s. The blue decoration is enamel (glass) that was put into parts of the pattern in paste form, and then fired to melt and harden it.

PEACE BE WITH YOU
The Hall of Supreme Harmony in Beijing's Forbidden City is guarded by this bronze statue of a turtle. Despite its rather fearsome appearance, the turtle was actually a symbol of peace.

GOLDEN FIREBIRDS
Chinese craftsmen fashioned these beautiful phoenix birds from thin sheets of delicate gold. The mythical Arabian phoenix was said to set fire to its nest and die, only to rise again from the ashes. During the Tang Dynasty, the phoenix became a symbol of the Chinese empress Wu Zetian, who came to power in AD660. It later came to be a more general symbol for all empresses.

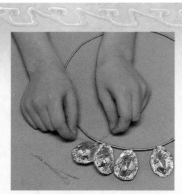

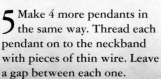

4 Glue the 2 pendant halves together, leaving one end open. Drop some rolled-up balls of foil into the opening. Seal the opening with glue.

5 Make 4 more pendants in the same way. Thread each pendant on to the neckband with pieces of thin wire. Leave a gap between each one.

People of all classes wore decorative jewelry in imperial China. The design of this necklace is based on the metal bell bracelets worn by Chinese children.

Chinese Firsts

WHEN YOU WALK DOWN a shopping street in any modern city, it is very difficult to avoid seeing some object that was invented in China long ago. Printed words on paper, silk scarves, umbrellas or locks and keys are all Chinese innovations. Over the centuries, Chinese ingenuity and technical skill have changed the world in which we live.

A seismoscope is a very useful instrument in an earthquake-prone country such as China. It was invented in AD132 by a Chinese scientist called Zhang Heng. It could record the direction of even a distant earth tremor. Another key invention was the magnetic compass. Around AD1–100, the Chinese discovered that lodestone (a type of iron ore) could be made to point north. They realized that they could magnetize needles to do the same. By about AD1000, they worked out the difference between true north and magnetic north and began using compasses to keep ships on course.

Gunpowder is another Chinese invention, from about AD850. At first it was used to blast rocks apart and to make fireworks. Later, it was used in warfare.

SHADE AND SHELTER
A Ching Dynasty woman uses an umbrella as a sunshade to protect her skin. The Chinese invented umbrellas about 1,600 years ago and they soon spread throughout the rest of Asia. Umbrellas became fashionable with both women and men and were regarded as a symbol of high rank.

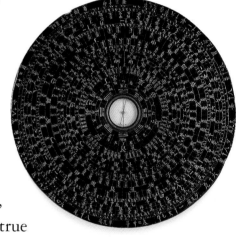

THE SAILOR'S FRIEND
The magnetic compass was invented in China around AD1–100. At first it was used as a planning aid to ensure new houses faced in a direction that was in harmony with nature. Later it was used to plot courses on long sea voyages.

MAKE A WHEELBARROW
You will need: thick card, ruler, pencil, scissors, compasses, ¼ in diameter balsa strips, glue and brush, paintbrush, paint (black and brown), water pot, 1½ in x ¼ in dowel, ¾ in diameter rubber washers (x4).

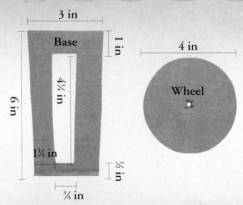

3 in

Base

1 in

4 in

Wheel

6 in

4¼ in

1¼ in

½ in

¾ in

Using the measurements above, draw the pieces on to thick card. Draw the wheel with the compasses. Cut out pieces with scissors.

1 Cut 2¾ in, 3 in and 10½ in (x2) balsa strips. Glue 2¼ in strip to short edge of base and 3 in strip to top edge. Glue 10½ in strips to side of base.

SU SONG'S MASTERPIECE

This fantastic machine is a clock tower that can tell the time, chime the hours and follow the movement of the planets around the Sun. It was designed by an official called Su Song, in the city of Kaifeng in AD1092. The machine uses a mechanism called an escapement, which controls and regulates the timing of the clock. The escapement mechanism was invented in the AD700s by a Chinese inventor called Yi Xing.

EARTHQUAKE WARNING

The decorative object shown above is the scientist Zhang Heng's seismoscope. When there was an earthquake, a ball was released from one of the dragons and fell into a frog's mouth. This showed the direction of the vibrations. According to records, in AD138 the instrument detected a earth tremor some 500km away.

ONE-WHEELED TRANSPORTATION

In about AD100, the Chinese invented the wheelbarrow. They then designed a model with a large central wheel that could bear great weights. This became a form of transportation, pushed along by muscle power.

The single wheelbarrow was used by farmers and gardeners. Traders wheeled their goods to market, then used the barrow as a stall. They sold a variety of goods, such as seeds, grain, plants and dried herbs.

2 Turn the base over. Cut two ¾ in x ½ in pieces of thick card. Make a small hole in the middle of each, for the wheel axle. Glue pieces to base.

3 Use compasses and a pencil to draw 1 circle around center of wheel and 1 close to the rim. Mark on spokes. Paint spaces between spokes black.

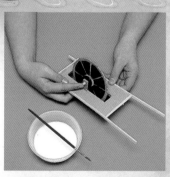

4 Paint the barrow, leave to dry. Cut two 2¾ in balsa strips with tapered ends to make legs, and paint them brown. When dry, glue to bottom of barrow.

5 Feed dowel axle between axle supports, via 2 washers, wheel, and 2 more washers. Dab glue on ends of axle to keep the wheel in place.

Extraordinary Chinese Engineering

THE ENGINEERING WONDER of ancient China was the Great Wall. It was known as *Wan Li Chang Cheng*, or the Wall of Ten Thousand *Li* (a unit of length). The Great Wall's main length was an incredible 3,977 mi. Work began on the wall in the 400s BC and lasted until the AD1500s. Its purpose was to protect China's borders from the fierce tribes who lived to the north. Despite this intention, Mongol invaders managed to breach its defenses time after time. However, the Great Wall did serve as a useful communications route. It also extended the Chinese Empire's control over a very long distance.

The Grand Canal is another engineering project that amazes us today. It was started in the 400s BC, but was mostly built during the Sui Dynasty (AD581–618). Its aim was to link the north of China with the rice-growing regions in the south via the Chang Jiang (Yangzi River). It is still in use and runs northwards from Hangzhou to Beijing, a distance of 1,115 mi. Other great engineering feats were made by Chinese mining engineers, who were already digging deep mine shafts with drainage and ventilation systems in about 160BC.

LIFE IN THE SALT MINES
Workers busily excavate and purify salt from an underground mine. Inside a tower (*shown bottom left*) you can see workers using a pulley to raise baskets of mined salt. The picture comes from a relief (raised carving) found inside a Han Dynasty tomb in the province of Sichuan.

MINING ENGINEERING
A Ching Dynasty official tours an open-cast coal mine in the 1800s. China has rich natural resources and may have been the first country in the world to mine coal to burn as a fuel. Coal was probably discovered in about 200BC in what is now Jiangxi province. Other mines extracted metals and valuable minerals needed for the great empire. In the Han Dynasty, engineers invented methods of drilling boreholes to extract brine (salty water) from the ground. They also used derricks (rigid frame-works) to support iron drills – over 1,800 years before engineers in other parts of the world.

HARD LABOR

Chinese peasants use their spades to dig roads instead of fields. Imperial China produced its great building and engineering works without the machines we rely on today. For big projects, work forces could number hundreds of thousands. Dangerous working conditions and a harsh climate killed many laborers.

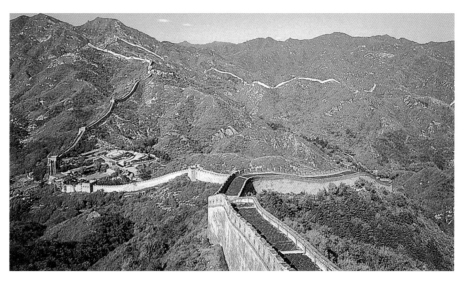

BUILDING THE WALL

The Great Wall snakes over mountain ridges at Badaling, to the northwest of Beijing. The Great Wall and Grand Canal were built by millions of workers. All men aged between 23 and 56 were called up to work on them for one month each year. Only noblemen and civil servants were exempt.

A GRAND OPENING

This painting from the 1700s imagines the Sui emperor Yangdi opening the first stage of the Grand Canal. Most of the work on this massive engineering project was carried out from AD605–609. A road was also built along the route. The transportation network built up during the Sui Dynasty (AD561–618) enabled food and other supplies to be moved easily from one part of the empire to another.

THE CITY OF SIX THOUSAND BRIDGES

The reports about China supposedly made by Marco Polo in the 1200s described 6,000 bridges in the city of Suzhou. The Baodai Bridge (*shown above*) is one of them. It has 53 arches and was built between AD618 and AD906 to run across the Grand Canal.

Chinese Science

From the Chinese Empire's earliest days, scholars published studies on medicine, astronomy and mathematics. The Chinese system of medicine had a similar aim to that of Daoist teachings, in that it attempted to make the body work harmoniously. The effects of all kinds of herbs, plants and animal parts were studied and then used to produce medicines. Acupuncture, which involves piercing the body with fine needles, was practiced from about 2700BC. It is believed to release blocked channels of energy and so relieve pain.

The Chinese were also excellent mathematicians, and from 300BC they used a decimal system of counting based on tens. They may have invented the abacus, an early form of calculator, as well. In about 3000BC, Chinese astronomers produced a detailed chart of the heavens carved in stone. Later, they were the first to make observations of sunspots and exploding stars.

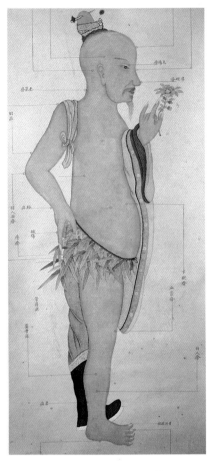

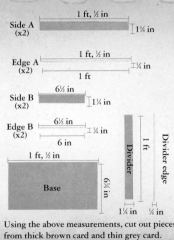

NEW ILLS, OLD REMEDIES
A pharmacist weighs out a traditional medicine. Hundreds of medicines used in China today go back to ancient times. Many are herbal remedies later proved to work by scientists. Doctors are still researching their uses. Other traditional medicines are of less certain value, but are still popular purchases at street stalls.

PRICKING POINTS
Acupuncturists used charts to show exactly where to position their needles. The vital *qi* (energy) is thought to flow through the body along 12 lines called meridians. The health of the patient is judged by taking their pulse. Chinese acupuncture is practiced all over the world today.

MAKE AN ABACUS

You will need: thick and thin card, ruler, pencil, scissors, wood glue and brush, masking tape, self-drying clay, cutting board, modeling tool, 1 ft x ¼ in dowel (x11), paintbrush, water pot, brown paint.

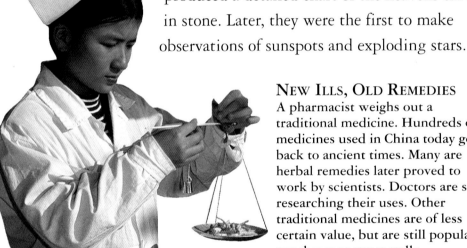

Side A (x2) — 1 ft, ½ in — 1¼ in
Edge A (x2) — 1 ft, ½ in — 1 ft — ¼ in
Side B (x2) — 6½ in — 1¼ in
Edge B (x2) — 6½ in — 6 in — ¼ in
Base — 1 ft, ½ in — 6¼ in
Divider — 1 ft — 1¼ in
Divider edge — 1 ft — ¼ in

Using the above measurements, cut out pieces from thick brown card and thin grey card. (pieces not shown to scale).

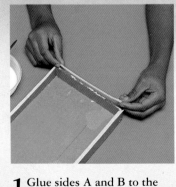

1 Glue sides A and B to the base. Hold the edges with masking tape until dry. Then glue edges A and B to the tops of the sides, as shown.

2 Roll the clay into a ¾ in diameter sausage. Cut it into 77 small, flat beads. Make a hole through the center of each bead with a dowel.

A STREET DOCTOR PEDDLES HIS WARES
This European view of Chinese medicine dates from 1843. It shows snakes and all sorts of unusual potions being sold on the streets. The doctor is telling the crowd of miraculous cures.

NATURAL HEALTH
Roots, seeds, leaves and flowers have been used in Chinese medicine for over 2,000 years. Today, nine out of ten Chinese medicines are herbal remedies. The Chinese yam is used to treat exhaustion. Ginseng root is used to help treat dizzy spells, while mulberry wood is said to lower blood pressure.

Chinese yam

ginseng root

BURNING CURES
A country doctor treats a patient with traditional techniques during the Song Dynasty. Chinese doctors relieved pain by heating parts of the body with the burning leaves of a plant called moxa (mugwort). The process is called moxibustion.

The abacus is an ancient counting frame that acts as a simple but very effective calculator. Using an abacus, Chinese mathematicians and merchants could carry out very difficult calculations quickly and easily.

3 Make 11 evenly spaced holes in the divider. Edge one side with thin card. Thread a dowel through each hole. Paint all of the abacus parts. Leave to dry.

4 Thread 7 beads on to each dowel rod – 2 on the upper side of the divider, 5 on the lower. Carefully fit the beads and rods into the main frame.

5 Each upper bead on the abacus equals 5 lower beads in the same column. Each lower bead is worth 10 of the lower beads in the column to its right.

6 Here is a simple sum. To calculate 5+3, first move down one upper bead (worth 5). Then move 3 lower beads in the same column up (each worth 1).

Specialist Crafts in Japan

FROM ANCIENT TIMES, the finest quality craftsmanship was important in Japan. Although paper was invented in China, in 105AD, when the Japanese started papermaking 500 years later, they raised it to a craft of the highest level. Different papers were made into both luxury and everyday objects – from wall-screens and lanterns to clothes, windows and partitions in houses.

Working with wood was another Japanese speciality. Doorways, pillars and roofs on most large Japanese buildings, such as temples and palaces, were elaborately carved or painted, or even gilded. Inside, beams and pillars were made from strong tree trunks, floors were laid with polished wooden strips, and sliding screens had fine wooden frames. A display of woodworking skill in a building demonstrated the owner's wealth and power. However, some smaller wooden buildings were left deliberately plain, allowing the quality of the materials and craftsmanship, and the elegance of the design, to speak for themselves.

WOODEN STATUES
This statue of a Buddhist god was carved between AD800 and 900. Many powerful sculptures were inspired by religion at this time.

SCREENS WITH SCENES
Screens were movable works of art as well as providing privacy and protection from drafts. This screen of the 1700s shows Portuguese merchants and missionaries listening to Japanese musicians.

ORIGAMI BOX
You will need: a square of origami paper (6 in x 6 in), clean and even folding surface.

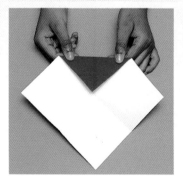

1 Place your paper on a flat surface. Fold it horizontally across the center. Next fold it vertically across the center and unfold.

2 Carefully fold each corner to the center point as shown. Unfold each corner crease before starting to make the next one.

3 Using the creases, fold all the corners back into the center. Now fold each side ¾ in from the edge to make a crease and then unfold.

GRAND PILLARS

This row of red wooden pillars supports a heavy, ornate roof. It is part of the Meiji Shrine in Tokyo. Red (or cinnabar) was the traditional Japanese color for shrines and royal palaces.

HOLY LIGHTS

Lamps of pleated paper were often hung outside Shinto shrines. They were painted with the names of people who had donated money to the shrines.

PAPER ART

Papermaking and calligraphy (beautiful writing) were two very important craft forms in Japan. These people have everything they need to decorate scrolls. Fine paper showed off a letter-writer's elegance and good taste.

PAPER RANGE

For artists such as the painter of this picture, Ando Hiroshige (1797–1858), the choice of paper was as important as the painting itself. The Japanese developed a great range of specialist papers.

Making boxes is a specialist craft in Japan. Boxes were used for storing all sorts of possessions.

4 Carefully unfold two opposite side panels. Your origami box should now look like the structure shown in the picture above.

5 Following the crease marks you have already made, turn in the side panels to make walls, as shown in the picture. Turn the origami round 90°.

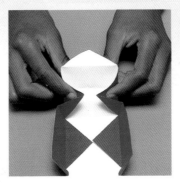

6 Use your fingers to push the corners of the third side in, as shown. Use the existing crease lines as a guide. Raise the box slightly and fold the wall over.

7 Next, carefully repeat step 6 to construct the final wall. You could try making another origami box to perfect your technique.

Fine Celtic Crafts

THE CELTS WERE A PROUD people to whom appearance was important. Beautiful objects carried important messages about their owner's wealth and power. From written descriptions of their clothes, we know that the Celts were skilled weavers and dyers. While the Roman invaders of their lands wore togas and tunics, the Celtic men were wearing trousers. Celtic craftworkers made finely worked jewelry, applying their expert glassmaking, enamelling and metalworking skills. They were also excellent potters. Some decorations have magic or religious meanings to protect people from harm or to inspire warriors setting off to war.

We do not know much about the craftworkers themselves. They may have been free and independent or the skilled slaves of wealthy families. However, toward the end of the Celtic period, around AD60, many craftworkers worked in *oppida* (Roman fortified towns) instead of in country villages.

SMOOTH AND SHAPELY
Tall, graceful vases with smoothly curving sides were a speciality of Celtic potters working in France. They date mostly from the La Tène era (450–50BC). Pots like these were produced on a potter's wheel. They were prestige goods, produced for wealthy or noble families.

ANGULAR ART
During the Hallstatt era (750–450BC), Celtic potters decorated their wares with spiky, angular designs like the patterns on this pottery dish. After about 500BC, when compasses were introduced into Celtic lands from countries near the Mediterranean Sea, designs based on curves and circles began to replace patterns made up of angles and straight lines.

MAKE A TORC

You will need: board, modeling clay, ruler, string, scissors, Elmer's glue and brush, gold or bronze paint, paintbrush.

1 On the board, roll out two lengths of modeling clay, as shown. Each length should be approximately 2 ft long and about ½ in thick.

2 Keeping the two lengths of clay on the board, pleat them together. Leave about 2 in of the clay unpleated at either end, as shown.

3 Make loops out of the free ends by joining them together. Dampen the ends with a little water to help join the clay if necessary.

PRECIOUS BOX

This gold and silver box was made in Scotland and was designed to hold Christian holy relics. It was associated with the Irish monk and Christian missionary St. Columba. After the saint's death, it was kept as a lucky talisman (charm), and carried into battle by Scottish armies.

GLASS JEWELS

Glass was made from salt, crushed limestone and sand, and colored by adding powdered minerals. Craftworkers melted and twisted different colored strands together to make jewel-like beads. Glass paste was applied to metalwork and fired to bond it to the metal in a process called enameling.

manganese

glass

cobalt

lead

MAKING WAVES

The sides of this pot are decorated with a molded pattern of overlapping waves. The pot has survived unbroken from the La Tène era over 2,000 years ago. It was found in France and it is made from fired clay. Celtic potters built elaborate kilns to fire (bake) their pots at high temperatures.

ELEGANT ENAMEL

This bronze plaque is decorated with red and yellow enamel. It was made in southern Britain around 50BC and was designed to be worn on a horse's harness.

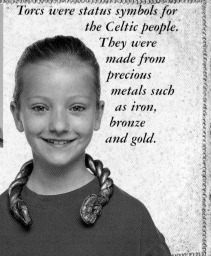

4 With the ruler, measure an opening between the two looped ends. The ends should be about 3½ in apart so that the torc fits easily around your neck.

5 When the torc is semi-dry, cut two pieces of string about 3 in long. Use the string to decorate the torc's looped ends. Glue the string in place.

6 Allow the clay to dry completely. When it is hard, cover all the clay and string with gold or bronze paint. Leave to dry.

Torcs were status symbols for the Celtic people. They were made from precious metals such as iron, bronze and gold.

Celtic Metalworkers

CELTIC METALWORKERS EXCELLED in several different techniques. They were among the most important people in Celtic society because they made many of the items that Celtic people valued most, from bronze and iron swords to beautiful gold jewelry. Patterns and techniques invented in one part of the Celtic world were copied and quickly spread to other parts. Designs were sketched onto the back of the metal, then gently hammered from beneath to create raised patterns. The technique was called repoussé (pushed out).

It took several years to learn all the necessary skills, and metalworkers probably began their training very young. They extracted iron from raw nuggets or lumps of ore in a very hot fire, and then forged the red-hot metal into shape. The Celts were the first to shoe horses, and invented seamless iron rims for chariot wheels to strengthen them.

PUSHED OUT DESIGN

This bronze shieldboss was made between about 200BC and 10BC. A boss is the metal plate that was fixed to the center of a shield to protect the hand of the person holding it. The raised pattern was created by pressing out the design in the thin covering sheet of metal from behind. The technique is called repoussé (pushed out).

TOOLS OF THE TRADE

Many bronze items, such as this horse's bit (below) and harness-ring (far left), were made by pouring molten metal into clay molds, then leaving it to cool and become solid. You can also see fragments of the clay molds, and the little crucible used for melting the bronze (top left).

MAKE A MIRROR

You will need: pair of compasses, pencil, ruler, stiff gold mirror card, scissors, tracing paper, pen, modeling clay, board, gold paint, paintbrush, Elmer's glue.

1 With the compasses, draw a circle 8¾ in wide on to gold card. Cut out. Use this circle as a template to draw a second circle on to gold card.

2 Cut out the second gold circle. Draw another circle on tracing paper. Fold the piece of tracing paper in two and draw a Celtic pattern in pencil.

3 Lay the tracing paper on to one of the circles. Trace the pattern on to half of the gold circle, then turn the paper over and repeat. Go over the pattern with a pen.

FROM EARTH AND SEA

The most valuable materials for metalworking were difficult and sometimes dangerous to find. Silver ore was dug from mines underground, or from veins in rocks on the surface. Miners searched for nuggets of gold in gravel at the bottom of fast-flowing streams. Swimmers and divers hunted for coral that grew on little reefs in the Mediterranean Sea.

bronze ore

coral

gold nuggets

BANDS OF GOLD

The Celts of the Hallstatt era (750–450BC) liked to wear bold, dramatic jewelry, such as the armband and ankle rings shown here. They were found in a tomb in central France. Both the armband and the ankle rings were made of sheets of pure gold and twisted gold wire which were carefully hammered and soldered together.

DELICATE DESIGN

This clothing toggle was created using the lost wax method of casting. The shape of the piece was modeled in beeswax, then the fine details were added. The wax model was covered with a thick layer of clay. Then the clay-covered model was heated, and the wax ran out. Finally, molten gold was poured into the space where the wax had been.

TOOLS OF THE TRADE

These little bone spatulas (knives for scooping and spreading) were used by metalworkers to add fine details to the surface of wax models when casting bronze objects using the lost wax process.

The bronze on a Celtic mirror would have polished up so that the owner could see his or her reflection in it.

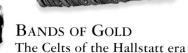

4 Roll out several snakes of modeling clay and sculpt them into a handle, as shown here. The handle should be about 6 in long and 3½ in wide.

5 Leave the modeling clay to dry. Then paint one side of the handle with gold paint. Leave to dry, then turn over and paint the other side.

6 Stick the two pieces of mirror card together, white side to white side. Glue the handle on to one side of the mirror.

Viking Crafts

SNARL OF THE DRAGON

This masterpiece of wood carving and metalwork is a dragon-head post. It is from the Oseberg ship burial in Norway and dates from about 850. Its patterns include monsters known as "gripping beasts."

IN EVERY VIKING HOME, people turned their hand to craftwork. The men made and repaired tools and weapons. They carved walrus ivory and wood during long winter evenings. The women made woolen cloth. They washed and combed the wool and then placed it on a long stick called a distaff. The wool was pulled out and spun into yarn on a whirling stick called a spindle. The yarn was woven on a loom, a large upright frame. Blacksmiths' furnaces roared and hammers clanged against anvils as the metal was shaped and reshaped.

Professional craftworkers worked gold, silver, bronze and pewter – a mixture of tin and lead. They made fine jewelry from amber and from a glassy black stone called jet. Beautiful objects were carved from antlers and ivory from the tusks of walruses. Homes, and later churches, had beautiful wood carvings. Patterns included swirling loops and knots, and birds and animals interlaced with writhing snakes and strange monsters.

SILVER SWIRLS

Can you see a snake and a beast in the design of this silver brooch? The Vikings were very fond of silver and collected hoards of coins, ornaments, silver ingots and jewelry from their raids.

MAKE A SILVER BRACELET

You will need: tape measure, self-drying clay, board, scissors, white cord or string, modeling tool, silver acrylic paint, paintbrush, water pot.

1 Measure your wrist with the tape measure to see how big your bracelet should be. Allow room for it to pass over your hand, but not fall off.

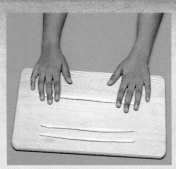

2 Roll the clay between the palms of your hand. Make three snakes that are just longer than your wrist measurement. Try to make them of equal thickness.

3 Lay out the three snakes on the board in a fan shape. Cut two lengths of white cord, a bit longer than the snakes, and place them in between.

COLORS FOR CLOTH

Woolen cloth was dyed in bold colors from leaves, roots, bark and flowers. A wildflower called weld, or dyer's rocket, was used for its yellow dye. The root of the madder gave a red dye. Bright blue came from the leaves of woad plants.

woad   *madder*

THE SMITH AT WORK

This fine wood carving comes from a church in Urnes, Norway. It shows Regin the blacksmith forging a sword on an anvil, for the legendary hero Sigurd. The smith is using bellows to heat up the furnace. The skills of metal working were so important in ancient times that smiths were often seen as magical figures or gods.

TOOLS FROM THE FORGE

Viking blacksmiths used hammers for beating and shaping metal. Tongs were used for handling red-hot iron. Shears were for cutting metal sheets. The blacksmith made everything from nails and knives to farm tools.

Vikings liked to show off their wealth and rank by wearing expensive gold and silver jewelry.

4 While the clay is still soft, pleat the snakes of clay and the two cords together. Ask an adult to help if you are not sure how to make a pleat.

5 Trim each end of the pleat with a modeling tool. At each end, press the strands firmly together and secure with a small clay snake, as shown above.

6 Carefully curl the bracelet around so that it will fit neatly over your wrist, without joining the ends. Leave it in a safe place to harden and dry.

7 When the bracelet is completely dry, paint it silver. Cover the work surface if necessary. Leave the bracelet to dry again – then try it on!

Mesoamerican Time

DIFFERENT CULTURES and civilizations devised different ways of splitting the year into seasons. The Egyptian calendar, based on a 365-day year, was linked to the annual flooding of the River Nile. The Maya and Aztec peoples in Mesoamerica had three different calendars. One, based on a 260-day year, was probably based on the time a baby spends in the womb. It was divided into 13 cycles of 20 days each. The calendar followed by Mesoamerican farmers was based on the movements of the Sun, because the seasons made their crops grow. Its 360-day year was divided into 18 months of 20 days, with five extra days that were considered unlucky. Every 52 years, measured in modern time, the two calendars ended on the same day. For five days before the end of the 52 years, people feared the world might end. There was a third calendar, of 584 days, used for calculating festival days.

SUN STONE
This massive carving was made to display the Aztec view of creation. The Aztecs believed that the world had already been created and destroyed four times and that their Fifth World was also doomed.

STUDYING THE STARS
The Caracol was constructed as an observatory to study the sky. From there, Maya astronomers could observe the planet Venus, which was important in the Mesoamericans' measurement of time.

MAKE A SUN STONE
You will need: pencil, scissors, thick card, self-drying clay, modeling tool, board, rolling pin, masking tape, Elmer's glue, glue brush, water bowl, pencil, thin card, water-based paints, paintbrush, water pot.

1 Cut a circle about 10 in in diameter from thick card. Roll out the clay and cut out a circle, using the card as a guide. Place the clay circle on the card one.

2 With a modeling tool, mark a small circle in the center of the clay circle. Use a roll of masking tape as a guide. Do not cut through the clay.

3 Carve the Sun-god's eyes, mouth, teeth and earrings. You can use the real Aztec Sun stone, shown at the top left of this page, as a guide.

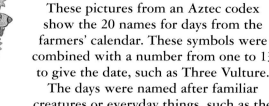

Names of Days

These pictures from an Aztec codex show the 20 names for days from the farmers' calendar. These symbols were combined with a number from one to 13 to give the date, such as Three Vulture. The days were named after familiar creatures or everyday things, such as the lizard or water. Each day also had its own god. Children were often named after the day on which they were born, a custom that still continues in some parts of Mexico up to the present day.

alligator	*wind*	*house*	*lizard*

serpent	*death's head*	*deer*	*rabbit*

Your finished Sun stone will not be as big as the original Aztec one. That measures 13 ft across and is the largest Aztec sculpture discovered so far.

water	*dog*	*monkey*	*grass*

reed	*jaguar*	*eagle*	*vulture*

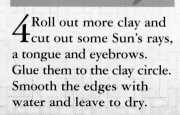   

motion	*flint knife*	*rain*	*flower*

4 Roll out more clay and cut out some Sun's rays, a tongue and eyebrows. Glue them to the clay circle. Smooth the edges with water and leave to dry.

5 Copy the 20 Aztec symbols (*above*) for days onto squares of thin card. The card squares should be no more than ¾ in x ¾ in. Cut out. Paint brown.

6 Cover the clay circle with a thin coat of dark brown paint. Leave it to dry. Then add a thin coat of white paint to make the circle look like stone.

7 Glue the card symbols evenly around the edge of the clay circle, as shown. Paint the Sun stone with a thin layer of Elmer's glue to seal and varnish it.

Practical Incas

Though the Incas were known for their fine work in precious metals, they also found practical solutions to more everyday needs. They built 14,914 mi of roads through the mountains of their empire. Their main buildings constructed from giant, many-sided blocks of stone all perfectly interlocking, were earthquake-proof.

Reeds and similar materials were woven from early prehistoric times into baskets and mats. Because there was no iron in the mountains, bone, stone and wood were carved into small items such as bowls, pins, spoons and figures. Pottery was made in Peru from about 2000BC, rather later than in the lands to the north and east, and revolutionized the production, storage, transportation and cooking of food. South American potters did not use a wheel to shape their pots, but built them up in layers from coils of clay. The coils were smoothed out by hand or with tools, marked or painted, dried in the sun and then baked hard.

Many of the pre-Incan civilizations of the Andes produced beautifully patterned pottery.

POLISHED WOOD
This fine black *kero* (drinking vessel) was made by an Inca craftsman. It is of carved and polished wood. Timber was always scarce in the Inca Empire, but wood was widely used to make plates and cups. Rearing up over the rim of the beaker is a fierce-looking big cat, perhaps a puma or a jaguar.

MODELED FROM CLAY
A fierce puma bares his teeth. He was made from pottery between AD500 and 800. The hole in his back was used to waft clouds of incense during religious ceremonies in the city of Tiwanaku, near Lake Titicaca.

A TIWANAKU POTTERY JAGUAR

You will need: chicken wire, wire-cutters, ruler, newspaper, scissors, Elmer's glue, masking tape, flour, water, card, paint, water pot, paintbrush.

1 Cut a rectangle of chicken wire about 5½ in long and 8 in wide. Carefully wrap it around to form a sausage shape. Close one end neatly.

2 Squeeze the other end of the sausage to form the jaguar's neck and head. Fold over the wire at the end to make a neat, round shape for his nose.

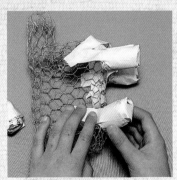

3 Make rolls of newspaper about 1 in long to form the jaguar's legs. Use strips of paper and glue to join them securely to the jaguar's body as shown.

PRETTY POLLY

This pottery jar, like many from Peru, comes with a handle and a spout. It is shaped and painted to look like a parrot and was made, perhaps 1,000 years before the Incas, by the Nazca potters of southern Peru.

WATER OF LIFE

This Inca bottle is carved with a figure inside a tower collecting water. No community could survive very long without a good supply of fresh water. Many pots, bottles and beakers from the South American civilizations are decorated with light-hearted scenes of everyday activities. They give us a vivid idea of how people used to live.

IN THE POTTER'S WORKSHOP

The potter needed a good supply of sticky clay and plenty of water. He also needed large supplies of firewood or dung for fuel. The potter would knead the clay until it was soft and workable. Sometimes he would mix in sand or crushed shells from the coast to help strengthen the clay. Colors for painting the pottery were made from plants and minerals.

shells *sand*

clay

The handle and spout design of your Tiwanaku jaguar is known as a stirrup pot, because the arrangement looks rather like the stirrup of a horse.

4 Mix the flour and water to a paste. Use it to glue a layer of newspaper strips all over the jaguar's body. Allow this layer to dry. You will need 3 layers.

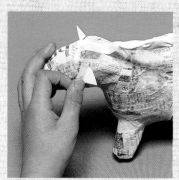

5 Cut ears from card. Fix on with masking tape. Tape on rolls of newspaper to make the handle, spout and tail as in the finished pot above.

6 Leave the model in a warm and airy place to dry. Then paint it all over with reddish brown paint. Allow the paint to dry completely.

7 Use black paint and a fine brush to decorate the jaguar as shown in the picture. When the paint is dry, varnish with Elmer's glue if you wish.

Inca Mining and Metalwork

THERE WERE RESOURCES of gold and silver in the Andes mountains and the Inca peoples became expert at working these precious metals into fabulous vessels, jewelry and life-sized figures of animals. Copper was mined for weapons. Metalworkers were highly respected members of Inca communities. A stone bowl that was discovered in the Andahuaylas Valley was nearly 3,500 years old. It contained metalworking equipment and finely beaten gold foil.

The Incas often referred to gold as "sweat of the Sun" and to silver as "tears of the Moon." These metals were sacred to the gods and also to the Inca emperor and empress. At the Temple of the Sun in the city of Cuzco, there was a whole garden made of gold and silver, with golden soil, golden stalks of corn and golden llamas. Copper was used by ordinary people. It was made into cheap jewelry, weapons and everyday tools.

The Incas' love of gold and silver eventually led to their downfall, for it was rumors of their fabulous wealth that lured the Spanish to invade the region.

A SICAN LORD
This ceremonial knife with its crescent-shaped blade is called a *tumi*. Its gold handle is made in the shape of a nobleman or ruler. He wears an elaborate headdress and large disks in his ears. It was made between 1100 and 1300. The knife is in the style of the Sican civilization, which grew up after the decline of the Moche civilization in the AD700s.

A CHIMÚ DOVE
Chimú goldsmiths, the best in the Empire, made this plump dove. When the Incas conquered Chimor in 1470, they forced many thousands of skilled craftsmen from the city of Chan Chan to resettle in the Cuzco area and continue their work.

A TUMI KNIFE
You will need: card, ruler, pencil, scissors, self-drying clay, cutting board, rolling pin, modeling and cutting tools, Elmer's glue, gold paint, paintbrush, water pot, blue metallic paper.

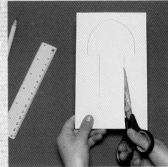

1 On card, draw a knife shape as shown and cut it out. The rectangular part should be 3½ in x 1¼ in. The rounded part is 2¾ in across and 1¾ in high.

2 Roll out a slab of clay about ½ in thick. Draw a *tumi* shape on it as shown. It should be 13 in long and measure 3½ in across the widest part at the top.

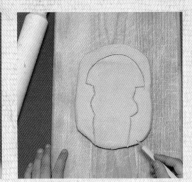

3 Use the cutting tool to cut around the shape you have drawn. Carefully take away the leftover clay. Make sure the edges are clean and smooth.

MINERAL WEALTH

To this day, the Andes are very rich in minerals. The Incas worked with gold, silver, platinum and copper. They knew how to make alloys, which are mixtures of different metals. Bronze was made by mixing copper and tin. However, unlike their Spanish conquerors, the Incas knew nothing of iron and steel. This put them at a disadvantage when fighting the Europeans.

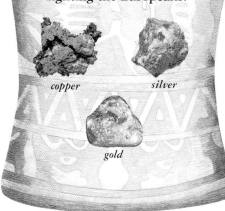

copper *silver*

gold

PANNING FOR GOLD

A boy laborer in modern Colombia pans for gold. Some Inca gold was mined, but large amounts also came from panning mountain rivers and streams in the Andes. The river bed was loosened with sticks, and then the water was sifted through shallow trays in search of any flecks of the precious metal that had been washed downstream.

INCA FIGURES

Small ritual figures of women and men from about 2½ in high were often made in the Inca period. They were hammered from sheets of silver and gold and were dressed in miniature versions of adult clothing. They have been found on mountain-top shrine sites in the south-central Andes, in carved stone boxes in Lake Titicaca and at important temples.

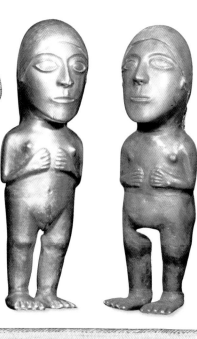

The Chimú gold and turquoise tumi *was used by priests at religious ceremonies. It may have been used to kill sacrifices.*

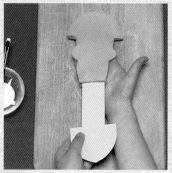

4 Cut a slot into the bottom edge of the clay shape. Lifting it carefully, slide the knife blade into the slot. Use glue to make the joint secure.

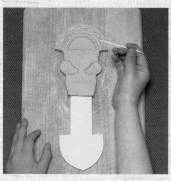

5 Use a modeling tool to mark the details of the god on to the clay. Look at the finished knife above to see how to do this. Leave everything to dry.

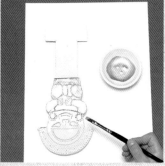

6 When the clay has hardened, paint the whole knife with gold paint. Leave it to dry completely before painting the other side as well.

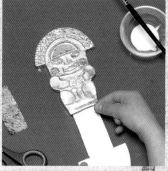

7 The original knife was decorated with turquoise. Glue small pieces of blue metallic paper on to the handle as shown in the picture above.

Inca Medicine and Magic

LIKE MOST PEOPLES in the world five hundred years ago, the Incas and their neighbors had some idea of science or medicine. However, curing people was believed to be chiefly a matter of religious rituals and magical spells. No doubt some of these did help people to feel better. Curing sick people was the job either of priests, or of the local healer or medicine man.

As in Europe at that time, Inca healers used fasting and blood-letting (allowing blood to flow from a cut) for many cures. They also tried blood transfusion (putting new blood into someone's body). They succeeded in this far earlier than doctors in other parts of the world, because peoples of the Andes shared the same blood group. The Incas could also set broken bones, amputate limbs, treat wounds and pull teeth. Medicines were made from herbs, roots, leaves and powders.

THE MEDICINE MAN
This Moche healer or priest, from about AD500, seems to be going into a trance and listening to the voices of spirits or gods. He may be trying to cure a sick patient, or he may be praying over the patient's dead body.

MAGIC DOLLS
Model figures like this one, made from cotton and reed, are often found in ancient graves in the Chancay River region. They are often called dolls, but it seems unlikely that they were ever used as toys. They were probably believed to have magical qualities. The Chancay people may have believed that the dolls helped the dead person in another world.

CARRYING COCA
Small bags like these were used for carrying medicines and herbs, especially coca. The leaves of the coca plant were widely used to stimulate the body and to kill pain. Coca is still widely grown in the Andes today. It is used to make the illegal drug cocaine.

MEDICINE BAG
You will need: scissors, cream calico fabric, pencil, ruler, paintbrush, water pot, acrylic or fabric paints, black, yellow, green and red wool, Elmer's glue, needle and thread, masking tape.

1 Cut two 8 in squares of fabric. Draw a pattern of stripes and diamonds on the fabric and use acrylic or fabric paints to color them.

2 For the tassels, cut about 10 pieces of wool 3 in long. Fold a piece of wool 6 in long in half. Loop it around each tassel as shown above.

3 Wind a matching piece of wool, 20 in long, around the end of the tassel. When you have finished, knot the wool and tuck the ends inside.

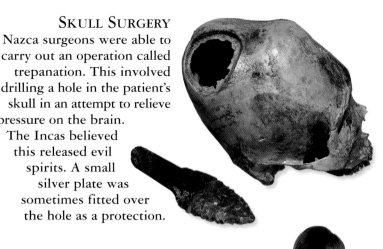

HERBAL REMEDIES

Drugs widely used in ancient Peru included the leaves of tobacco and coca plants. A yellow-flowered plant called calceolaria was used to cure infections. Cinchona bark produced quinine, a medicine we use today to treat malaria. That illness only arrived in South America after the Spanish conquest. However, quinine was used earlier to treat fevers. Suppliers of herbal medicines were known as *hampi kamayuq*.

cinchona tree   tobacco plant

SKULL SURGERY

Nazca surgeons were able to carry out an operation called trepanation. This involved drilling a hole in the patient's skull in an attempt to relieve pressure on the brain. The Incas believed this released evil spirits. A small silver plate was sometimes fitted over the hole as a protection.

Doctor on call! An Inca medicine chest took the form of a woven bag, carried on the shoulder.

A BAD OMEN

A comet shoots across the night sky. The Incas believed such sights would bring plague or disease in their wake. Other common causes of illness were believed to include witchcraft, evil spirits and a failure to please the gods. People tried to make themselves better by making offerings to the gods at *waq'as* (local shrines). Healers used charms or spells to keep their patients free from evil spirits.

4 Make nine tassels in all. Place them in groups of three along the bottom of the unpainted side of one of the pieces of fabric. Use glue to fix them in place.

5 Allow the glue to dry. Place the unpainted sides of the fabric pieces together. Sew around the edges as shown. Leave the top edge open.

6 Make a strap by pleating together strands of wool as shown. Cross each outer strand in turn over the middle strand. Tape will help keep the work steady.

7 Knot the ends of the strap firmly. Attach them to both sides of the top of the bag with glue. Make sure the glue is dry before you pick the bag up.

Calculations Inca-style

FORTUNES FROM THE STARS AND PLANETS
An Inca astrologer observes the position of the Sun. The Incas believed that careful watching of the stars and planets revealed their influence on our lives. For example, the star pattern or constellation that we call the Lyre was known to the Incas as the Llama. It was believed that it influenced llamas and those who herded them.

INCA MATHEMATICIANS used a decimal system, counting in tens. To help with their arithmetic, people placed pebbles or grains of corn in counting frames. These had up to twenty sections. *Quipu* strings were also used to record numbers. Strings were knotted to represent units, tens, hundreds, thousands or even tens of thousands.

The Incas worked out calendars of twelve months by observing the Sun, Moon and stars as they moved across the sky. They knew that these movements marked regular changes in the seasons. They used the calendar to tell them when to plant crops. Inca priests set up stone pillars outside the city of Cuzco to measure the movements of the Sun.

As in Europe at that time, astronomy, the study of the stars, was confused with astrology, which is the belief that the stars and planets influence human lives. Incas saw the night sky as being lit up by gods and mythical characters.

THE SUN STONE
A stone pillar called *Inti Watana* (Tethering Post of the Sun) stood at the eastern edge of the great square in Machu Picchu. It was like a giant sundial and the shadows it cast confirmed the movements of the Sun across the sky – a matter of great practical and religious importance.

A QUIPU
You will need: scissors, rope and string of various thicknesses, a 3 ft length of thick rope, paints, paintbrush, water pot.

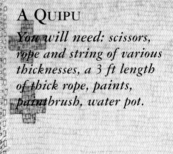

1 Cut the rope and string into about 15 lengths measuring from 8 in to 2 ft, 7½ in. Paint them in various bright colours. Leave them to dry completely.

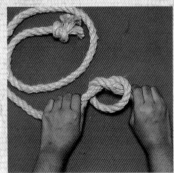

2 To make the top part of the *quipu*, take a piece of thick rope, about 3 ft long. Tie a knot in each end as shown in the picture above.

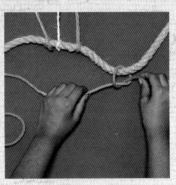

3 Next, take pieces of thinner rope or string of various lengths and colors. Tie them along the thicker rope, so that they all hang on the same side.

THE MILKY WAY

On dark nights, Inca priests looked for the band of stars that we call the Milky Way. They called it *Mayu* (Heavenly River) and used it to make calculations about seasons and weather conditions. In its darker spaces they saw the shadow of the Rain god Apu Illapu. The shape of the Milky Way was believed to mirror that of the Inca Empire.

SUN WATCH

The *Inti Watana* (Tethering Post of the Sun) at Machu Picchu was one of many Sun stones across the Empire. *Sukana* (stone pillars) near Cuzco showed midsummer and midwinter sun positions. The Sun god, Inti, was believed to live in the north and go south each summer.

KEEPERS OF THE QUIPU

Vast amounts of information could be stored on a *quipu*. A large one might have up to 2,000 cords. The *quipu* was rather like an Inca version of the computer, only the memory had to be provided by the operator's brain rather than a silicon chip. Learning the *quipu* code of colors, knots, and major and minor strings took many years. Expert operators were called *quipu-kamayuq*.

You have now designed a simple quipu. Can you imagine designing a system that would record the entire population of a town, their ages, the taxes they have paid and the taxes they owe? The Incas did just that!

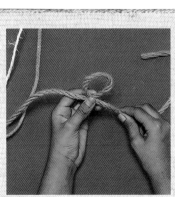

4 Tie knots in the thinner ropes or strings. One knot you might like to try begins by making a loop of rope as shown in the picture above.

5 Pass one end of the rope through the loop. Pull the rope taut but don't let go of the loop. Repeat this step until you have a long knot. Pull it tight.

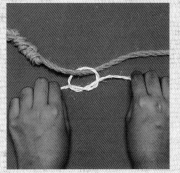

6 Make different sizes of knots on all the ropes or strings. Each knot could represent a family member, school lesson or other important detail.

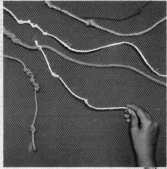

7 Add some more strings to the knotted strings. Your *quipu* may be seen by lots of people. Only you will know what the ropes, strings and knots mean!

Tribal Crafts in North America

NORTH AMERICAN INDIANS were expert craftsmen and women. Beautiful pots have been found dating back to around 1000BC. The people of the Southwest were renowned for their pottery. Black and white Mimbres bowls were known as burial pots because they were broken when their owner died and buried along with the body. Baskets and blankets were the other most important crafts. The ancient Anasazis were known as the basket-making culture because of the range of baskets they produced. Some were coiled so tightly they could hold water. The Apaches coiled large, flat baskets from willow and plant fiber, and the Paiutes made cone baskets, which were hung on their backs for collecting food. All North American Indians made use of the materials they had to hand such as wood, bark, shells, porcupine quills, feathers, bones, metals, hide and clay.

BASKET WEAVER
A native Arizona woman is creating a traditional coiled basket. It might be used for holding food or to wear on someone's head. Tlingit and Nootka tribes from the Northwest Coast were among those who wore cone-shaped basket hats.

POTTERY
Zuni people in the Southwest created beautiful pots such as this one. They used baskets as molds for the clay or coiled thin rolls of clay around in a spiral. Afterward, they smoothed out the surface with water. Birds and animals were favorite decorations.

DRILLING WALRUS TUSKS
An Inuit craftsman is working on a piece of ivory. He is using a drill to etch a pattern. The drill bit is kept firmly in place by his chin. This way, his hands are free to move the bow in a sawing action, pushing the drill point into the ivory.

MAKE A TANKARD
You will need: air-drying modeling clay, board, water in pot, pencil, ruler, cream or white and black poster paints or acrylic paints, fine and ordinary paintbrushes, non-toxic varnish.

1 Roll out a round slab of clay and press it into a flat circle with a diameter of about 4 in. Now, roll out two long sausage shapes of clay.

2 Slightly dampen the edges of the clay circle. Place one end of the clay sausage on the edge of the circle and coil it around. Carry on spiraling around.

3 Continue coiling with the other clay sausage. Then, use your dampened fingers to smooth the coils into a good tankard shape and smooth the outside.

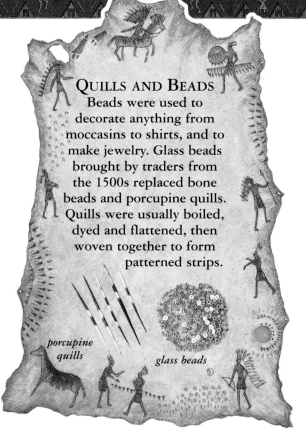

QUILLS AND BEADS

Beads were used to decorate anything from moccasins to shirts, and to make jewelry. Glass beads brought by traders from the 1500s replaced bone beads and porcupine quills. Quills were usually boiled, dyed and flattened, then woven together to form patterned strips.

porcupine quills

glass beads

TALKING BLANKET

It could take half a year for a Tlingit woman to make one of the famous Chilkat blankets. She wove cedar bark fiber and mountain goat wool with her fingers. The Tlingits said that if you knew how to listen, the blankets could talk.

FRUITS OF THE LOOM

Striped blankets were the speciality of Indians in the Southwest. This Hopi woman is using an upright loom made from poles. Pueblo people were the first North American Indians to weave like this.

Each tribe had its own pottery designs and colors. These geometric patterns were common in the Southwest.

4 Roll out another, small sausage shape of clay to make a handle. Dampen the ends and press it on to the clay pot in a handle shape. Leave to dry out.

5 Using a sharp pencil, mark out the design you want on your jug. You can follow the traditional indian pattern or make up your own.

6 Using poster paints or acrylic paints, color in the pattern on the mug. Use a fine-tipped brush to create the tiny checked patterns and thin lines.

7 When the paint is dry, coat your mug in one or two layers of nontoxic varnish using an ordinary paintbrush. This will protect it.

Glossary

A

abacus A wooden frame with beads on rods, used for calculating.

acupuncture The treatment of the body with fine needles to relieve pain or cure illness.

adze A woodcutting tool with a blade at right angles to the handle.

alloy A mixture of metals melted together to create a new metal that may be stronger or easier to work.

Anno Domini (AD) A system used to calculate dates after the supposed year of Christ's birth. AD dates are written before the date (e.g. AD521).

anvil A heavy iron block on which metal objects can be hammered into shape.

aqueduct A channel carrying water supplies. It may take the form of a bridge when it crosses a valley.

astronomy The scientific study of stars, planets and other heavenly bodies. In ancient times it was often mixed up with astrology, the belief that heavenly bodies shape our lives.

B

barter The exchange of goods, one for the other.

Before Christ (BC) A system used to calculate dates before the supposed year of Christ's birth. Dates are calculated in reverse (e.g. 200BC is longer ago than 1 BC). The letters BC follow the date (e.g.455BC).

bleeding The old medical practice of letting out blood that was thought to drain off poisons.

bow drill A tool used to drill holes in bone and shell, and also to generate heat to make fire.

brazier A metal stand for holding burning coals.

bronze A metal alloy, made by mixing copper with tin.

burin A chisel-like flint tool.

C

cast An object shaped by pouring metal or plastic into a mold and allowing it to set.

causeway A raised walkway.

civilization A society that makes advances in arts, sciences, law, technology and government.

clan A group of people related to each other by ancestry or marriage.

Confucianism The Western name for the teachings of the philosopher Kong Fuzi (Confucius), which call for social order and respect for one's family and ancestors.

Cro-Magnons The first modern humans to live in Europe.

cubit A unit of measurement, the length of a forearm.

D

distil The process of heating liquid to boiling point and collecting the condensed steam to make a purer liquid. Alcoholic spirits are made in this way.

dynasty A period of rule by the same royal family.

E

electrum A mixture of gold and silver, used for making coins.

empire An area including many different cities and countries and ruled by one person.

enamel A hard, colored glass-like substance, applied as a decorative or protective covering to metal or glass.

evolution A gradual change, maybe over thousands of years, during which the thing that is changing becomes more complex.

F

feud A long-standing quarrel, especially between two families.

firing The process of baking clay or glass paste in a kiln to harden it and make it waterproof.

G

geometric pattern A pattern made by lines, circles and triangles.

geometry A branch of mathematics concerning the measurements of lines, angles and surfaces.

gilding The process of applying a thin layer of gold, to metal or pottery, for example.

groma An instrument used by Roman surveyors to measure right angles and straight lines.

H

herbalism A method of healing people by using medicines made from plants.

Homo habilis (handy man) The first humans to make tools.

Homo sapiens (wise man) The species to which all modern humans and Neanderthals belong.

I

ice age One of a number of times in the Earth's history when large parts of the planet surface became covered with ice.

imperial Relating to the rule of an emperor or empress.

inlay To set or embed pieces of wood or metal in another material so that the surfaces are flat.

inscribed Lettering, pictures or patterns carved into stone or wood.

iron ore The rock that contains iron in its raw, natural form. The ore has to be crushed and then heated to release the metal.

ivory The hard, smooth tusks of elephants and walruses.

K

kaolin A fine white clay used in porcelain and papermaking.

kiln Industrial oven in which clay or enamel, for example, are fired.

L

lacquer A shiny varnish made from the sap of trees.

lodestone A type of magnetic iron ore, also called magnetite.

loom A frame or machine used for weaving cloth.

M

midwife Someone who provides care and advice for women, before and after childbirth.

migration The movement of people to other regions either permanently or at specific times of the year.

molten Something that is melted.

mosaic A craft in which tiny pieces of colorful stone, shell or glass are pieced together to make a design or picture to decorate floors, walls, tables and smaller objects.

N

nomad Person who moves from one area to another to find food or better land or to follow herds.

P

papyrus A tall reed that is used to make a kind of paper.

pewter An alloy or mixture of metals, made from tin and lead.

philosophy A Greek word meaning love of knowledge. Philosophy is the discipline of thinking about the meaning of life.

plumbline A weighted cord, held up to see if a wall or other construction is vertical.

politics The art and science of government (from *polis*, city state).

porcelain The finest quality of pottery. It was made with kaolin and baked at a high temperature.

pyramid A large monument with a square base, rising to a point or with steps up to a platform.

R

radiocarbon dating A scientific method of estimating the age of archaelogical objects.

relic Part of the body of a saint or martyr, or of some object connected with them, preserved as an object of respect and honor.

relief A sculpture in which a design is carved from a flat surface.

repoussé A metalworking technique that is used to create decorative raised patterns on a metal object.

S

seal A mark that is stamped or engraved on an object to show ownership or as proof of quality. The tool for making such a mark.

seismoscope An instrument that reacts to earthquakes and tremors.

smelt To heat rock to a high temperature in order to melt and extracting the metal within it.

slaves People who were not free but were owned by their masters.

soldered Something that is joined together with small pieces of melted metal.

spindle A whirling tool used to make fiber, such as wool, into yarn for weaving.

stylus A pointed tool, such as the one used to scratch words onto a wax tablet.

T

tapestry A cloth with a design sewn onto it.

terra cotta A composition of baked clay and sand used to make statues, figurines and pottery.

textile Any cloth that has been woven, such as silk or cotton.

treadwheel A wooden wheel turned by the feet of people, that was used to power mills or other machinery.

tribe A group of people who share a common language and way of life.

W

woad A blue dye extracted from a plant that ancient Britons used to decorate their bodies.

woodblock Hard wood into which a design has been carved, that is used for printing that design on to surfaces such as cloth or paper.

Index

A

abacus 40, 41
acupuncture 40
adzes 10
antlers 10-11, 48
aqueducts 18, 30
arches 31, 39
architecture 22, 23, 30
astrology 20, 26, 58
astronomy 5, 6, 7, 16, 17, 18, 21, 24, 40, 50, 58

B

basketmaking 12, 52, 60
blacksmiths 46, 48-49
boat-building 6, 7
bone 4, 10-11, 47, 52
bricks 15
bridges 31, 39
bronze 5, 18, 29, 32, 34, 35, 46, 47, 55

C

calendars 5, 16, 20, 50-51, 58
canals 6, 18, 38, 39
carbon-dating 14
casting 5, 19, 35, 46-47
ceramics 14-15
  see also pottery
childbirth 33
clay 5, 8, 12-3, 14-5, 19, 52
clinics 33
clocks 7, 21, 26, 37
cloth 4, 12, 18, 19, 48, 49
clothes 44
compasses 44
concrete 6, 30
copper 18, 54, 55
cotton 48
counting frames 40, 41, 58
cranes 30

D

decimal system 5, 20, 40, 58
decoration 7, 10, 11, 13, 14, 19, 34, 35, 44, 45, 46, 48, 52, 60, 61

doctors see medicine
drugs 56, 57

E

enameling 35, 44, 45
engineering 6, 7, 22-23, 24-25, 30-31, 38-39

F

farming 4, 12
flax 12, 48
flint 4, 8-9
forging 46, 48, 49
fuel 34

G

gearing systems 26
geometry 27
glassmaking 5, 44, 45
gold craft 5, 7, 19, 22, 25, 35, 45, 47, 54-55
gunpowder 7, 36

H

Hippocratic oath 28

I

iron 5, 34, 35, 46, 49
irrigation 18
ivory 10, 11, 48, 60

J

jewelry 5, 18, 45, 47, 48, 49, 54

K

keys 36

L

labor 18, 22, 24, 39
leatherwork 18
leveling, land 24
limestone 25, 31
linen 48
locks 36
looms 12

M

magic 20, 32, 44, 56
magnetic compasses 6, 36
marble 31

mathematics 5, 6, 7, 16, 20-21, 22, 26, 27, 40, 58-59
measuring 17, 20-21
medicine 16, 17, 20, 21, 28-29, 32-33, 40-41, 56-57
metalworking 5, 18, 19, 34-35, 44, 46-47, 48-49, 52, 54-55
mining 8, 35, 38, 47, 55

P

paper 6, 36, 42-43
pharmacy 32
philosophy 5, 26
plague 28
pottery 4, 5, 7, 12-13, 14-15, 18, 19, 44-45, 52-54, 60
printing 6, 7, 36
projects
  abacus 40-41
  Archimedes screw 26-27
  bracelet 48-49
  clay pot 12-13
  groma 30-31
  leg offering 28-29
  lion weights 16-17
  medicine bag 56-57
  mirror 46-47
  model ax 8-9
  necklace 34-35
  origami box 42-43
  painted plate 18-19
  pottery jaguar 52-53
  pyramid 24-25
  quipu 58-59
  sun stone 50-51
  tankard 60-61
  torc 44-45
  tumi knife 54-55
  water clock 20-21
  water pot 14-15
  wheelbarrow 36-37
public buildings 30, 31
pyramids 5, 22-25

Q

quarrying 24, 31
quipu strings 58, 59

R

remedies, herbal 5, 16, 17, 21, 28, 29, 32, 33, 40, 41, 56, 57
road systems 6, 30, 52

S

seismoscopes 36, 37
silkmaking 6, 36
silvercraft 5, 7, 34, 45, 47, 48, 54-55
sky maps 17, 21, 40
slate 31
slot machines 26
smelting 34, 35
soapstone 13
sphinx 23
spinning 12, 19, 48
steam power 7, 26
stone 4, 8-9, 22, 24, 30, 31, 52
surgery 20, 29, 32, 57

T

terra cotta 14-15
tools 4, 5, 8-9, 10-11, 18, 34, 35, 46, 47, 48, 49, 60
trading 7, 8
transportation 6-7, 39
tufa 31

U

umbrellas 36

W

walls 31, 38, 39
water control 18
water lifters 27
weapons 6, 8-9, 18, 19, 34, 35, 46
weaving 12, 13, 19, 44, 48, 60, 61
weights and measures 17
wheelbarrows 4, 37
wheels 4, 5, 14, 37
woodcraft 6, 10, 42, 43, 48, 52
wool 48

ARCTIC LANDS

ARCTIC LANDS

VIKING LANDS

North
America

CELTI
LAND

AZTEC & MAYA
EMPIRES

Gulf of
Mexico

Atlantic Ocean

Caribbean
Sea

Central
America
(Mesoamerica)

Pacific Ocean

INCA
EMPIRE

Andes Mountains

South
America

Cape Horn